Water Quality and Fish Health Management

भाकृअनुप
ICAR

Indian Council of Agricultural Research (ICAR)

Dr. Balasaheb Sawant Konkan Krishi Vidyapeeth (DBSKKV)

NAHEP

National Agricultural Higher Education Project-Institutional Development Plan (NAHEP-IDP)

Chulalongkorn University
Bangkok, Thailand

Water Quality and Fish Health Management

Compiled by
Prashant S. Bodake
Dean & Director of Instruction, DBSKKV
Co-ordinator & Principal Investigator
NAHEP-IDP, DBSKKV-Dapoli, Maharashtra

Nopadon Pirarat
Professor
Chulalongkorn University, Bangkok, Thailand

Authors
B.R. Chavan
A.S. Pawase
P.E. Shingare
Dr. Balasaheb Sawant Konkan Agriculture University, Dapoli, Maharashtra

Manoj Kamble
Priyanka Kuwar
Seema Medhe
Aranya Ponpornpisit
Channarong Rodkhum
Chulalongkorn University, Bangkok, Thailand

Indian Council of Agricultural Research (ICAR)

Dr. Balasaheb Sawant Konkan Krishi Vidyapeeth (DBSKKV)

National Agricultural Higher Education Project-Institutional Development Plan (NAHEP-IDP)

Chulalongkorn University Bangkok, Thailand

Citation: National Agriculture Higher Education Project Institutional Development Plan, DBSKKV, Dapoli & ICAR, New Delhi

NIPA® GENX ELECTRONIC RESOURCES & SOLUTIONS P. LTD.

101,103, Vikas Surya Plaza, CU Block
L.S.C.Market, Pitam Pura, New Delhi-110 034
Ph : +91 11 27341616, 27341717, 27341718
E-mail: newindiapublishingagency@gmail.com
Website: www.nipabooks.com

For customer assistance, please contact
Phone: + 91-11-27 34 17 17
Fax: + 91-11-27 34 16 16

ISBN: 978-93-58876-84-0

Composed and Designed by NIPA®.

DR. BALASAHEB SAWANT KONKAN KRISHI VIDYAPEETH
Dapoli, Dist. Ratnagiri - 415712
Tel. No.: (0)02358-282064, **Fax No.**: 02358-282074, **Mob.**: 9422556565
E-mail : vcdbskkv-mh@gov.in, ipmbhave@gmail.com, **website** : www.dbskkv.org

Dr. Sanjay G. Bhave
IDP Leader
Vice Chancellor, DBSKKV

Message

Fisheries and Aquaculture is an age-old practice and a way of life of some sectors of the society in India. In recent past, the traditional art of aquaculture became a well-defined science with research support and development of near-exacting technologies. It assumes more importance, as the fish production from capture fisheries has reached the limits and the natural resources are in danger of over-exploitation. In India, aquaculture plays an important role for the betterment of rural economy through better use of land and water resources. The fish productivity of rural ponds increased from about 50 Kg ha-1 year-1 in early 1960s to about 3.2 tons ha-1 year-1 at present as a national average. To meet the ever-growing population, India needs to produce 10 million tons of fish to provide 10 kg per capita per year as one of the animal protein sources in diet. The total inland fish production is 7.21 million metric tons during 2015-16 and almost 80% is contributed by freshwater aquaculture. Thus, the inland aquaculture sector would be the only sector to increase the fish production to meet mostly the domestic demand and nutritional security of the country. However, the presently available resources are not fully utilized to enhance the inland fish production.

World Bank funded; Indian Council of Agricultural Research New Delhi National Agricultural Higher Education Project implemented at Dr. Balasaheb Sawant Konkan Agricultural University Dapoli, Ratnagiri under this project "International Training on Fisheries and Aquaculture 2023" (ITAF2023) organized by Chulalongkorn University, Thailand from 22 September 2023 to 20 October, 2023. Book on Water Quality and Fish Health Management compiled which is the outcome of International Training on Aquaculture and Fisheries (ITAF-2023). This book will be serving as the innovative academic guide line for the students & faculty and farmers to enhance fish production and provide food security and improve their livelihood.

The efforts taken by Dr. Prashant Bodake, Dr. B.R.Chavan, Dr. A.S. Pawase, Dr. P. E. Shingare, Dr. Manoj Kamble, Dr. Seema Medhe, Dr. Nopadon Pirarat, Dr. Aranya Ponpornpisit and Dr. Channarong Rodkhum and all Editorial committee members of College of Fisheries, Shirgaon, Ratnagiri under the guidance of Dr. Prashant Bodake, PI, NAHEP-IDP project and Dean & Director of Instruction, DBSKKV, Dapoli are appreciated. This Book on Water Quality and Fish Health Management will be beneficial to every student, faculty members who is involved in teaching, research for empowerment of rural communities.

(Sanjay G. Bhave)
IDP Leader
Vice Chancellor, DBSKKV

NATIONAL AGRICULTURE HIGHER EDUCATION PROJECT INSTITUTIONAL DEVELOPMENT PLAN
DR. BALASAHEB SAWANT KONKAN KRISHI VIDYAPEETH,
DAPOLI, DIST. RATNAGIRI
Mob. No.: 9420413255 / 8275201130
Email.: (P)prameghash@gmail.com (O) hodagrodapoli@gmail.com (O) piidpdbskkv@gmail.com (O)nodalcelldbskkv@gmail.com

Dr. P. S. Bodake
Dean & Director of Instruction
Coordinator, Principal Investigator, NAHEP-IDP

Message

Education is a process where one learns subjects in specific field for deeper understanding and consequent advancing of its skylines beyond the current levels. University education is principally meant to be in this direction. However, it would not be possible for all to pursue many topics of frontier science to an understandable level. It is therefore imperative that the educationists should mold the students not only lead them to excellence in the specific subject area, but also to horn their skills and knowledge in the new developments in the associated subject areas which may not be covered under the class room teachings. For the young students, who are growing up through the system of professional education, the formative years of the person would be completely covered during the span of the educational period.

Under the World Bank funded; ICAR- National Agricultural Higher Education Project- Institutional Development Plan (NAHEP-IDP) one month "International training on Fisheries and Aquaculture 2023" (ITAF 2023) was organized by Chulalongkorn University, Thailand in collaboration with Dr. Balasaheb Sawant Konkan Agricultural University Dapoli, Ratnagiri from 22 September 2023 to 20 October, 2023. The said one month training program includes eminent scientist lectures, demonstrations, hands-on and field visits of aquaculture farms, mariculture farms and feed mill plants. I got an opportunity to participate in the various field visits and delivered concluding remark in closing ceremony of the training program. The said one-month international training program was concluded with course evaluation, post training examination, feedback from participated students and faculty members. The outcome of this one-month training program is prepared this book entitled "Water Quality and Fish Health Management" The information's presented in the book is in-depth.

The efforts taken by Dr. B. R. Chavan, Dr. A.S. Pawase, Dr. P. E. Shingare, Dr. Manoj Kamble, Dr. Seema Medhe, Dr. Nopadon Pirarat, Dr. Aranya Ponpornpisit and Dr. Channarong Rodkhum and all Editorial committee members, Amrendra Kumar, Viraj Daunde, Devaranjani VK, Rohit Apte, Prachi Dongarkar, Gargi Palekar and Swapnil Jadhav of College of Fisheries, Shirgaon, Ratnagiri. I do hope that this edition will generate inspiration among students and provided a good platform for bringing out their abilities.

Thank you.

(P. S. Bodake)
Dean & Director of Instruction
Coordinator, Principal Investigator, NAHEP-IDP

Dr. Nopadon Pirarat
Associate Professor and Director
ITAF 2023
Chulalongkorn University, Bangkok, Thailand

Message

The book titled "Water Quality and Fish Health Management" is a comprehensive guide to the International Training in Aquaculture and Fisheries 2023, hosted by the Faculty of Veterinary Science at Chulalongkorn University, Bangkok, Thailand from September 20 to October 20, 2021. This meticulously crafted guide aims to provide an in-depth overview of the diverse range of practical sessions, and enriching field visits that constitute this transformative program. Designed to equip participants with cutting-edge knowledge and practical skills in the dynamic fields of aquaculture and fisheries, this training endeavor stands at the forefront of global initiatives to advance sustainable practices and meet the growing demand for aquatic products.

In today's world, where issues of food security, environmental sustainability, and economic development are of paramount concern, the significance of aquaculture and fisheries cannot be overstated. This training program is tailored to address the multifaceted challenges and opportunities within these sectors, ensuring that participants are equipped with the latest insights and best practices to navigate this evolving landscape.

The book "Water Quality and Fish Health Management" serves as a roadmap, outlining the various components that comprise the training program. Furthermore, immersive field visits to renowned aquaculture facilities, fish farms, and fisheries management sites offer a unique opportunity to witness these principles in action and gain valuable insights from industry experts.

I extend my deepest gratitude to the Indian Council of Agricultural Research, ICAR- National Agricultural Higher Education Project- Institutional Development

Plan (NAHEP-IDP), New Delhi, India, the authority of Dr. Balasaheb Sawant Konkan Krishi Vidyapeeth, Dapoli, Ratnagiri, MS, India and the Ratchadapisek Sompoch Fund, Chulalongkorn University, Bangkok, Thailand, for their support in making this program possible. I also want to acknowledge ITAF coordinator, Dr. Manoj Kamble, Dr. Seema Medhe and all other committee members who have tirelessly worked to organize this training.

We hope that this book serves as a valuable companion throughout your journey in this International Training in Aquaculture and Fisheries. May it inspire you to embrace innovation, champion sustainability, and forge meaningful connections within this dynamic and vital field.

Wishing you a fruitful and enlightening experience.

Nopadon Pirarat
Associate Professor and Director
ITAF 2023
Chulalongkorn University, Bangkok, Thailand

Dr. B.R. Chavan
Professor & Head,
Department of Fisheries Engineering
College of Fisheries, (Dr. BSKKV) Shirgaon, Ratnagiri

Preface

It gives us immense pleasures in preparing this book entitled "Water Quality and Fish Health Management" is a outcome of international training program on Fisheries and Aquaculture 2023" (ITAF2023) organized by Chulalongkorn University, Thailand in collaboration with Dr. Balasaheb Sawant Konkan Agricultural University Dapoli, Ratnagiri from 22 September 2023 to 20 October, 2023 with financial support from World bank funded project ICAR- National Agricultural Higher Education Project- Institutional Development Plan (NAHEP-IDP).

All efforts are made to present the correct and authentic information in this book gathered. The first section of this book contains hands-on examination of finfish and shellfish Health, includes: Health examination of finfish, necropsy technique and sample collection technique, Microbiology diagnosis, Bacterial isolation and identification, Preparation of subculture plate, Health Examination of Shrimp, Molecular technology. In the second section of this contains water quality of aquaculture pond, it includes, Estimation of pH of aquaculture pond water, Analysis of Ammonia nitrogen, nitrite nitrogen and dissolved oxygen of aquaculture pond water. In the third section of this book contains, small scale aquaculture systems employed in Thailand. However, some mistake has been made indicated it be brought to our notice which will be taking care of in our next edition.

The authors are acknowledging to the authority of Dr. Balasaheb Sawant Konkan Krishi Vidyapeeth, Dapoli, Ratnagiri and ICAR- National Agricultural Higher Education Project- Institutional Development Plan (NAHEP-IDP) for financial assistance for preparation of this book. The authors are acknowledging to the authority of Chulalongkorn University, Thailand for organizing one moth international training program on Fisheries and Aquaculture 2023" (ITAF2023). The authors are also thankful to the NIPA publication, New Delhi for timely completing the assignment.

B.R. Chavan

Contents

Health Examination of Finfish and Shellfish

1

Finfish Health Examination, Necropsy Technique and Sample Collection Technique

Introduction

The study of fish health is crucial for understanding aquatic ecosystems and ensuring the well-being of aquatic species. This report provides a detailed account of a hands-on activity involving necropsy and health examination of fish. Through this practical experience, we aimed to gain insights into the common ailments affecting fish populations and develop skills in identifying pathological conditions.

The necropsy process involves the post-mortem examination of fish specimens, enabling us to observe and analyze internal organs, tissues, and external features. By combining practical techniques with theoretical knowledge, we seek to contribute to the broader understanding of fish health and disease.

This report will detail the methodology employed. The hands-on nature of this activity not only enhances our understanding of aquatic pathology but also equips us with valuable skills for monitoring and safeguarding the health of fish populations in diverse environments.

Objective

1. To understand the procedure of blood sample collection.
2. Carry out internal organ examination of fish.

Methodology

A) Blood collection from live fish

Protocol

1. Anesthetize the fish using few drops of clove oil
2. Hold the fish when the movement is completely ceased.
3. Collect blood from caudal vein by using sterilized syringe.
4. Blood can be collected by cardiac puncture also.
5. Collected blood should be stored in lithium heparin tube immediately.

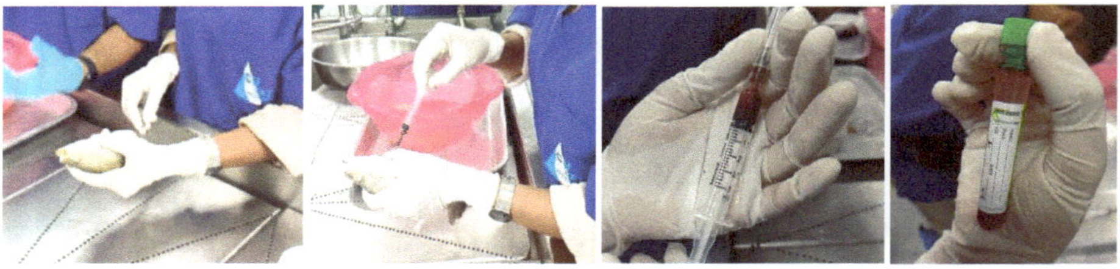

Fig. 1. Various steps involved in blood sample collecion

B) Preparation of blood smear slide

Protocol

1. Take sterilized glass slide ,place on uniform surface.(rough portion of slide at the end should face upward. Name it from bottom side.)
2. Add 1 drop of blood at one end of slide near rough portion.
3. Place the cover slip at 45° angle and gently wipe towards backward a little and then to another side.
4. Spread blood in thin layer and leave it to dry.

C) Dissection of fish and internal organ examination

Protocol

1. Use alcohol spray to disinfect the fish.
2. Use scissor of proper size.
3. Start cutting from anus towards pelvic fin.
4. Then cut abdominal part to expose the internal organs.
5. Use forceps to separate each part for easy identification.

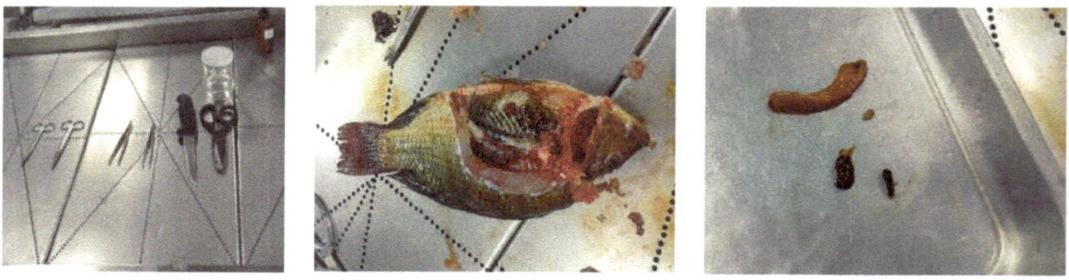

Fig. 2. Dissection activity

D) Preparation of primary culture plate:

Protocol

1. Burn the inoculation loop until it is red
2. Wait for 30 seconds to cool.
3. Swab the preferred internal organ with the loop without changing it´s original condition.
4. Streak the bacteria on culture media of sheep blood.
5. Incubate at 28°C for 24 hrs.

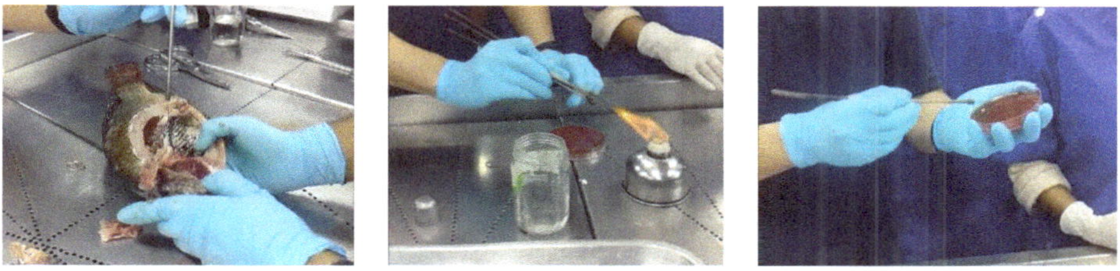

Fig. 3. Steps involved in preparation of primary culture plate

2

Microbiology Diagnosis, Bacterial Isolation and Identification

Introduction

In this microbiology report, we delve into the intricate realm of bacterial isolation and identification, exploring the fundamental processes that unlock the mysteries of microbial worlds. From the methods employed for isolating bacterial strains to the intricate techniques of identification, this report aims to provide a comprehensive understanding of the crucial role microbiology plays in unraveling the secrets of these microscopic organisms. Join us on a journey through the principles and applications that form the backbone of bacterial isolation and identification in the fascinating field of microbiology.

Objectives

1. To carry out gram staining to distinguish between gram positive and gram negative bacteria.
2. To understand procedures of various microbiological test to identify bacteria.

A) Gram staining

Protocol

1. Take 1 drop of water on the slide.
2. Pick a colony by a sterile loop and gently mix in water drop.
3. Add crystal violet (primary stain) for 1 min, then wash with water.
4. Add lugol´s iodine (mordant) for 1 min, then wash with water.
5. Directly wash with acetone(decolorizer).
6. Add safranin (counter stain) and wait for 30 sec, wash with water.
7. Wait until it is dry and observe under microscope.

Note: Gram-positive bacteria – purple
Gram negative bacteria – red

Fig. 4. Gram staining activity

B) Oxidase test

Protocol

1. Take filter paper which contain oxidase reagent.
2. Pick colony of bacteria using sterilized loop.
3. Strike on the paper and observe.

Note: If deep purple came, then test will be positive.

C) Catalase test

Protocol

1. Take clean slide
2. Add 1 drop of catalase reagent (hydrogen peroxide).
3. Isolate bacterial colony, gently put on catalase drop.
4. Observe bubble formation.

Note: Bubble formation means that the isolated organism is catalase positive.

Results

Vibrio spp. – catalase positive, oxidase positive.

Edwardsiella spp.- catalase positive, oxidase negative.

Aeromonas spp. - catalase positive, oxidase positive.

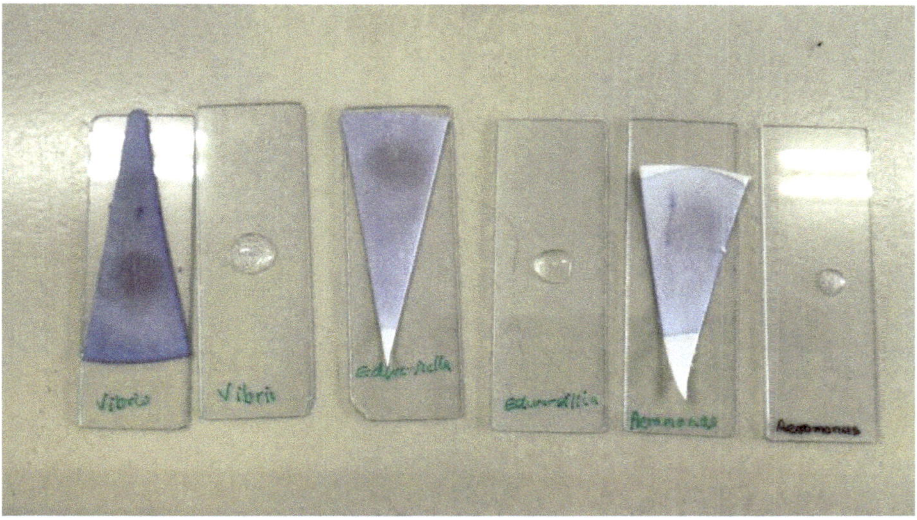

Fig. 5. Results of oxidase and catalase test

3

Health Examination of Shrimp

We performed a detailed hands-on practical on 'Shrimp Health Examination' at the Pathology Necroscopy room on 29th September 2023. Several activities regarding the health examination of shrimp were performed. It included, the collection of blood from an in-between portion of pereiopods and pleopods, also making blood smears on a glass slide, staining, and observation of blood cells under a microscope. Collection of samples from hepatopancreas for microbial culture to know the presence of any pathogen present. Collection of body part samples like gill, pleopods, and telson to observe the presence of external parasites. The species used health examination Whiteleg shrimp (*Litopennaenus vannamei*).

Activity-1 Procedure of Shrimp blood collection, smear preparation & staining

- Collect the live and healthy shrimp and take it on the necropsy table.
- Hold the shrimp on the ventral side, and clean the blood collection site with help of alcohol to disinfect it.
- Take the 1ml. syringe in another hand and hold carefully.
- Insert the needle in the middle portion of the periopods and pleopods gently and take out the blood carefully of about 0.1 to 0.2 ml. quantity and remove the needle.
- Now transfer the very little quantity of collected blood on a glass slide.
- Take a coverslip, hold at about 45p angle, and spread the blood throughout the slide immediately as shrimp blood coagulates in a very short time. (from a rough side portion of the slide to the other end)
- Let the smear completely dry under the air.
- Now for staining, dip the slide in methanol solution for 5 seconds and remove.
- Then, dip the slide for 5 times in Solution A (red color solution)
- Finally, immediately dip the slide in crystal violet solution (Solution B) 5 times and then let the slide dry completely.

- After drying the slide, keep it under a microscope to see the blood cells like haematocyte, nucleus, granules, etc., and observe for any abnormalities in blood cells to detect the presence of any blood disease.

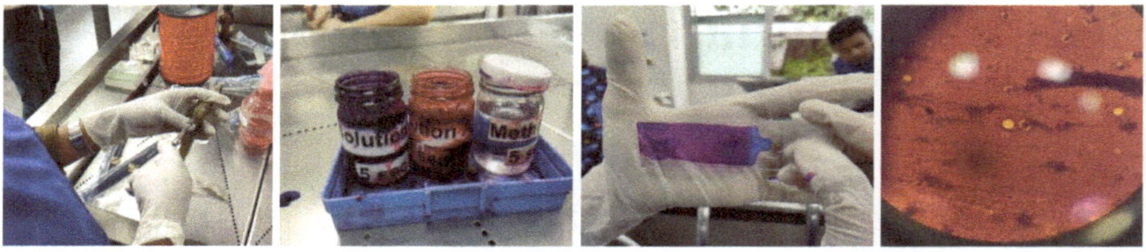

Fig. 6. Blood sample collection from shrimp and staining

Activity-2 Procedure of collection of samples for microbial analysis from Haepatopancreas

- Take the live and healthy shrimp on the necropsy table.
- Now take sterilized forceps and scissor.
- Clean the carapace portion properly with the help of alcohol and wipe it with the help of cotton to disinfect the portion.
- Gently cut open the portion from the carapace with the help of scissor till we see the hepatopancreas globule.
- Care should be taken that no portion of the hand should touch the opened portion to avoid contamination.
- Take the inoculation loop and heat it on flame until it turns bright red. Wait for a few seconds until it cools.

  Insert the loop inside the hepatopancreas globule gently puncture it and take out the loop.
- Now, take the appropriate media required for culture like 'Sheep blood'.
- Make strikes on the media plate in a zig-zag manner. Ensure to make about 4 such streaks around the plate and make sure to connect all the streaks to let the bacterial colonies grow in all streaks.
- After around 24 hours, check under a microscope to identify the pathogen species by following bacterial species identification protocol.

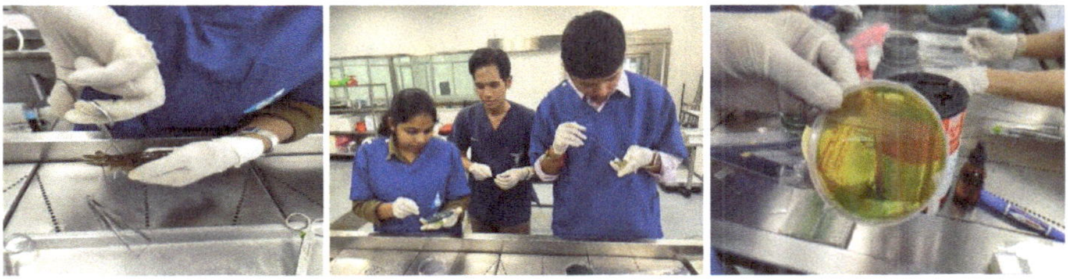

Fig. 7. Sample collection activity

Activity-3 Procedure of collection of body parts sample to identify the presence of external parasites

- Take the same shrimp used for the collection of samples from hepatopancreas.
- Take the sterilized forceps and scissor.
- Keep ready the glass slide and cover slips, and put small three drops of water on it.
- Now, with the help of scissor, cut open the gill portion cut the small tip of it, and put it on a glass slide, into the drop of water.
- In a similar manner, cut the tip of the pleopod and telson and put it in a drop of water placed on the slide.
- Place coverslips properly on it.
- Keep the slide under the microscope and check properly if any external parasite is found on small portions of the gill, pleopods, and telson, The Presence of external parasites can also be confirmed by noticing any damage on these parts.

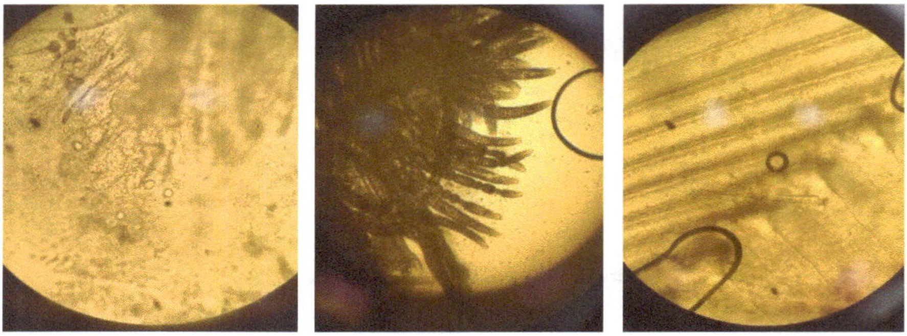

Fig. 8. sample collection and observing under microscope

4

Preparation of Subculture Plate

Introduction

Subculture is key skill in bacteriological laboratory allowing us to colony propagate a pure culture in single colony. Pure cultures are essential for characterization such as identification using biochemical test or antimicrobial susceptibility testing.

Protocol

1. Label your fresh blood agar plate.
2. Select a single colony to subculture (colonies grows by clonal expansion; each colony should be a pure strain.)
3. Use the colony you have just picked to make the 1st streak on your fresh plate.
4. Your 1st steak should cover 1/5 of the plate.
5. A sterile loop is used to make the 2nd streak
6. In each streak, the concentration of bacteria is diluted.
7. Be careful not to touch your loop to the primary streak.
8. 4th streak should cover more area without touching other 3 streaks.
9. Incubate at 28°C for 24 hrs.

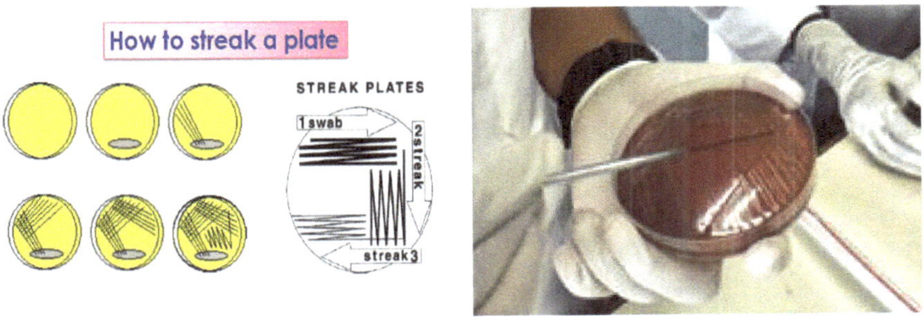

Fig. 9. Method of steaking a plate

5

Molecular Technology

Introduction

In the ever-evolving realm of fisheries and aquaculture, the infusion of molecular technology has become a beacon of hope, transforming the way we diagnose diseases and identify pathogens. Imagine it as a kind of genetic detective work that not only helps us understand the health of aquatic life but also empowers us to protect it. It all begins with the delicate dance of DNA extraction, a bit like teasing out the secrets locked within the cells of aquatic organisms. This biological treasure map, once unveiled, guides us through the labyrinth of diseases. Enter the PCR protocol for Carp Edema Virus (CEV) detection—a sort of molecular magnifying glass that allows us to zoom in on specific genetic footprints of trouble. It's like deciphering a genetic code, helping us pinpoint potential issues swiftly. And then, there's gel electrophoresis, a visual maestro that orchestrates the separation of DNA fragments, revealing the unique genetic melodies of pathogens. Together, these techniques form a compassionate narrative, painting a picture of health and resilience for our aquatic ecosystems, where science becomes a guardian, ensuring the well-being and sustainability of our underwater companions.

Objectives

The objective was to equip participants with practical skills and knowledge in the application of molecular technology for disease diagnosis and pathogen identification in the context of fisheries and aquaculture.

To gain proficiency in the essential molecular techniques, enabling to effectively contribute to the advancement of disease management and biosecurity measures in the fisheries and aquaculture industry.

1. DNA extraction

Protocol

The DNA extraction protocol involves several steps to isolate DNA from fish tissue. Initially, 5-10 mg of fish tissue is carefully weighed and placed into a 1.5 ml microcentrifuge tube. Following this, 100 μl of Buffer R1 is added to the sample for resuspension. To denature proteins, 180 µl of Buffer R2 and 20 µl of Proteinase K are added, and the mixture is incubated at 65°C for 20 minutes. Homogenization is achieved by adding 600 µl of Buffer G and incubating at the same temperature. Precipitation of DNA is facilitated by adding 200 µl of absolute ethanol.

The next step involves loading the sample onto a column and centrifuging it at 10,000 × g for 1 minute, after which the flow-through is discarded. The column is then washed with 650 µl of wash buffer and centrifuged again at 10,000 × g for 1 minute. To ensure thorough drying, the column is centrifuged once more at the same speed for 1 minute. Subsequently, the elution step is carried out by transferring the column to a new microcentrifuge tube and adding 50 µl of Elution buffer. The mixture is allowed to stand for 2 minutes.

For the final step, DNA is collected by centrifuging at 10,000 × g for 1 minute. The extracted DNA can then be stored at a temperature of -20°C for future use. This comprehensive protocol enables the successful isolation of DNA from fish tissue, making it ready for downstream molecular biology applications.

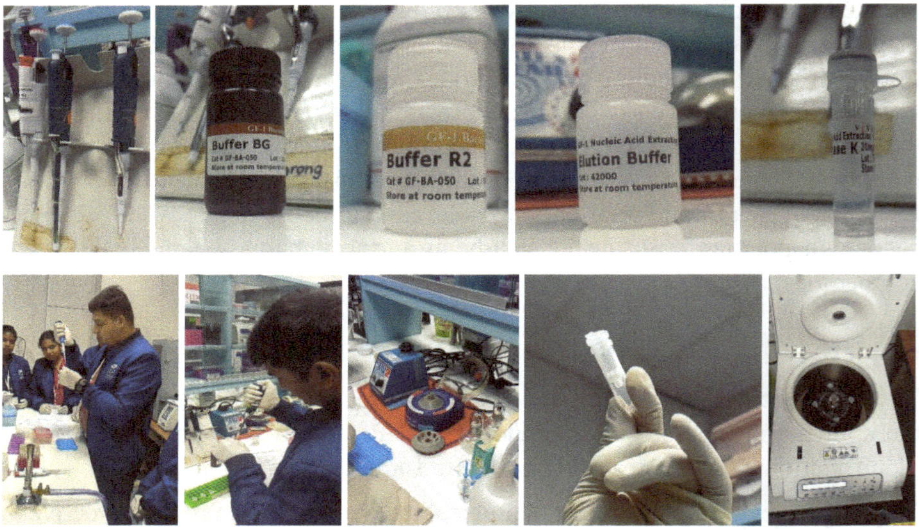

Fig. 10. Material and methods

2. PCR protocol for Carp Edema Virus (CEV) detection

The PCR protocol for Carp Edema Virus (CEV) detection utilizes the CEFAS end-point PCR assay, a method developed by the Centre for Environment, Fisheries and Aquaculture Science as described in the study by Matras et al. in 2017. This highly sensitive and specific assay is designed to identify the presence of CEV in fish samples. It involves the amplification of target DNA sequences using specific primers, followed by a post-amplification analysis to determine the presence or absence of the virus. This protocol is a valuable tool for the surveillance and early detection of CEV in aquatic environments, contributing to the effective management and control of this viral pathogen in carp populations.

Primer name	Sequence (5'-3')	Product size
CEV-for-B	ATG GAG TAT CCA AAG TAC TTA G	528bp
CEV-rev-J	CTC TTC ACT ATt GTG ACT TTG	

PCR reaction preparation

2× Master mix	10 μl
10 μM CEV-for-B	0.8 μl
10 μM CEV-rev	0.8 μl
Ultrapure water	4.4 μl
50 ng/μl DNA templet	4 μl
Total	20 μl

Amplification program

Program	Temperature	Time	Cycle No.
Initial denaturation	94 ^{0}C	2 min	1
Denaturation	94 ^{0}C	20 sec	35
Annealing	55 ^{0}C	10 sec	
Extension	72 ^{0}C	50 sec	
Final extension	72 ^{0}C	5 min	1

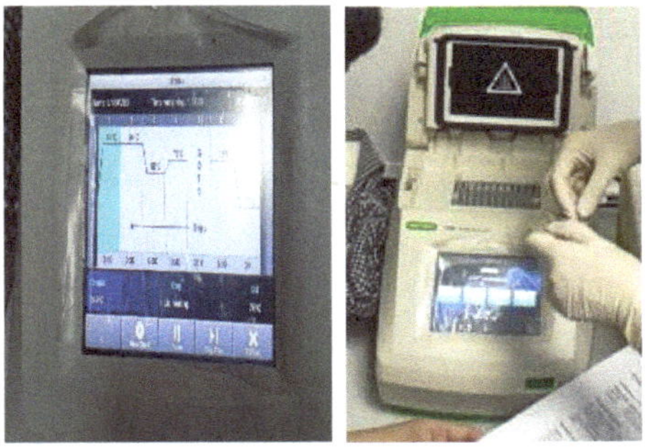

Fig. 11. Polymerase Chain Reaction (PCR)

3. Gel electrophoresis

Gel electrophoresis is a widely used laboratory technique employed to separate and analyze DNA fragments based on their size and charge. The protocol for gel electrophoresis typically involves casting a gel made of agarose or polyacrylamide, which serves as a molecular sieve, allowing smaller DNA molecules to migrate faster than larger ones through an electric field. To visualize and estimate the size of the DNA fragments, a base pair ladder, consisting of known-sized DNA fragments, is loaded into one of the wells. This ladder acts as a reference, helping researchers determine the approximate size of the sample DNA. Additionally, positive and negative probes may be used to detect specific DNA sequences or proteins within the gel. The positive probe is designed to bind to the target molecule, generating a signal, while the negative probe is used to verify the absence of the molecule in question. Altogether, these components are crucial in the gel electrophoresis process, facilitating precise DNA analysis and molecular detection.

Reagent preparation

10 × TBE buffer

Tris base	108 g
Boric acid	55 g
EDTA	7.4 g
Distilled water	800 ml

Once dissolved, top up the solution to 1 L

1× TBE buffer

10× TBE buffer	100 ml
Distilled water	900 ml

1.5% agarose gel

Agarose gel	1.5 g
1× TBE buffer	100 ml
DNA staining solution	5 μl

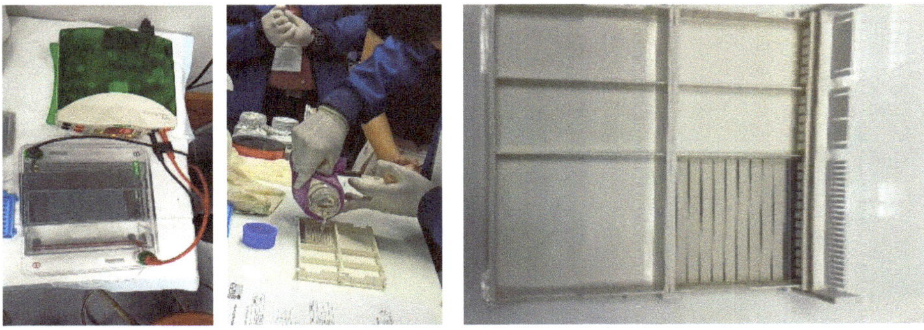

Fig. 12. Gel electrophoresis: Material and Method

Estimation of Water Quality Parameters of Aquaculture Pond

6

Estimation of Water Quality Parameters of Aquaculture Pond

Introduction

Water is the chief factor in fish farming. Water carries oxygen, eliminate metabolic waste products, through its physical and chemical composition, determine the success of production. The major constraint is that the waste organic matter released into water.

Measurement of pH, temperature, conductivity, salinity, transparency, turbidity and solids are important parameters to play role in water quality. pH in water changes during the day and night in response to photo-synthesis which is supported by the sun light intensity. Temperature is a physical parameter which effected on fish growth as resulting of many bio-chemical processes. It changes with the time depending on air temperature surrounding site. In water body, it also varies with the depth of water. Conductivity in water is an ability to conduct an electrical current. The greater of ions in the water provided higher the conductivity. If the ions proportionally differ among the water, it is proportional to the salinity and may be different. The depth of visibility of an underwater object such as the Secchi disk affords a measure of transparency. In fish ponds, plankton is usually the major source of turbidity, so Secchi disk visibility has frequently been used to estimate plankton density. The term of turbidity refers to the presence of suspended matter in water which interferes with the passage of light through water. Many substances can impart turbidity to water but plankton and suspended soil particles are the most common sources of turbidity in pond aquaculture. Solids consist including dissolved organic matter, particulate organic matter, dissolved inorganic substances except gases, the carbon dioxide contained in bicarbonate and particulate inorganic substances. The total suspends solids and the settle able solids in the water discharged from ponds indicated the pollution strength of the effluent.

Objectives

1. To get experience measurements of pH, temperature, Do, ammonia, nitrite and nitrate.
2. To learn the techniques for estimation of water quality parameters

Methodology

1. pH

Protocol

1. Rinse the test vial with water sample and fill to 5 ml mark.
2. Add 4 drops of solution A and mix.
3. Place the vial on color chart and match with the value.

Note : pH range : 6.5-9.3

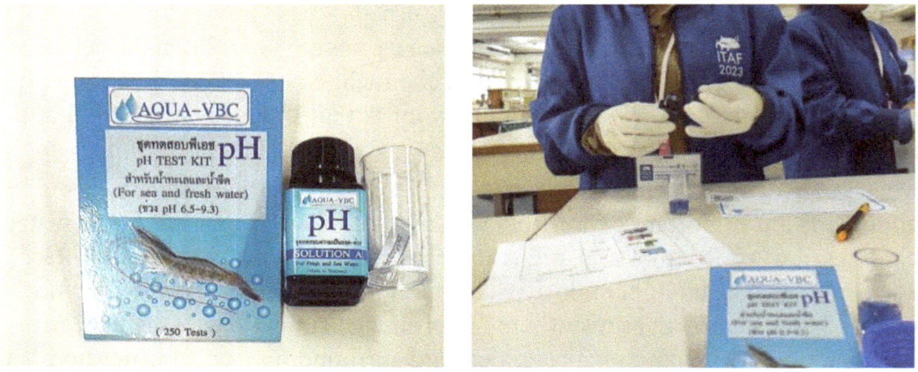

Fig. 13. Test kit used for estimating pH.

7

Analysis of Ammonia Nitrogen of Aquaculture Pond Water

Introduction

Ammonia is the major end product in the breakdown of proteins in fish. Fish digest the protein in their feed and excrete ammonia through their gills and in their faces. The amount of ammonia excreted by fish varies with the amount of feed put into the pond or culture system – increasing as feeding rates increase. Ammonia also enters the pond from bacterial decomposition of organic matter such as uneaten feed or dead algae and aquatic plants.

Ammonia is extremely toxic and even relatively low levels pose a threat to fish health. Ammonia is produced by fish and all other animals, including ourselves, as part of normal metabolism. Such is the toxicity, that most animals immediately convert it to a less harmful substance, usually urea, and excrete it in urine. Ammonia levels in a new set-up can be minimized by only introducing a few new fish at a time and allowing the nitrifiers to 'catch up' with the increase in ammonia before putting more fish in. There are bacterial cultures available that are supposed to speed up filter maturation, but even with these there will still be a time lag before nitrification cuts in

Objective

1. To get experience analysis of ammonia at the end of the practice.
2. To estimate ammonia nitrogen value of the pond water.

Protocol

1. Rinse the test vial with water sample and fill to 5 ml mark.
2. Add 5 drops of solution A mix and leave for 1 minute.
3. Add 1 spoon of powder B mix to dissolve.
4. Add 5 drops of solution C, mix and leave for 5 min.
5. With pH, compare with the standard color chart to obtain the ammonia concentration.

Note: The chemicals used are corrosive. If accidently contact, rinse with water several times.

Fig. 14. Test kit used for estimating ammonia

8

Analysis of Nitrite Nitrogen

Introduction

Ammonia, nitrite and nitrate are considerable in the water quality management because they can be produced a toxic depending to their form and concentration. The solubilities of nitrogen contain in water at different temperatures and pressures. Nitrogen is a basic growth substance and an important element in aquatic ecosystems as well as in the pond water. It exists variously forms in water and manly classified in organic and inorganic. The first stage in nitrification is ammonia being converted to less toxic nitrite (NO2-) by *Nitrosomonas sp*. As the population of *Nitrosomonas* grows, the ammonia level starts to drop. However, this is usually followed by an increase in nitrite levels, which persists until the population of a different bacterium, *Nitrobacter sp.* reaches optimum levels. *Nitrobacter* convert nitrite to nitrate which is usually considered harmless at levels less than 50 mg/litre.

Objectives

1. To get experience analysis of nitrite at the end of the practice.
2. To estimate nitrite nitrogen value of the pond water.

Protocol

1. Rinse the vials several times with water sample and fill to 5 ml mark.
2. Add 5 drops of solution A, mix and wait for 1 minute.
3. Add 1 spoon of powder B,mix to dissolve.
4. After 3-4 min, compare the color with standard color chart.

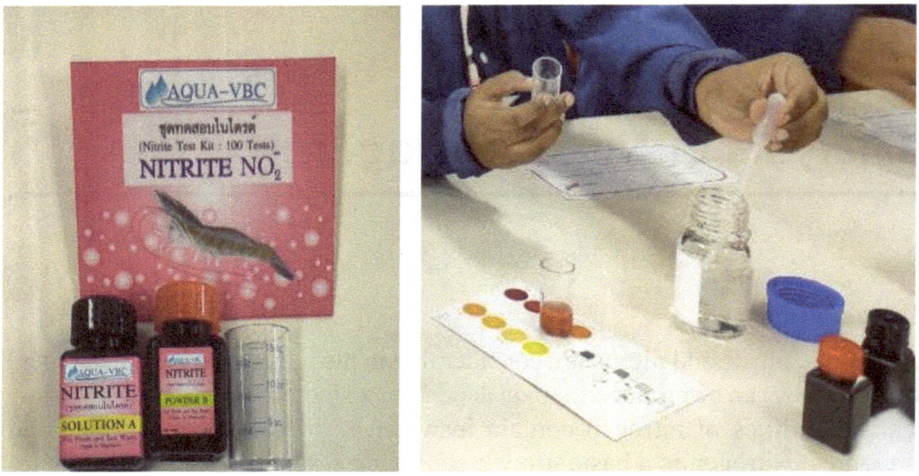

Fig. 15. Test kit used for estimating nitrite

9

Analysis of Dissolved Oxygen

Introduction

The minimum dissolved oxygen (DO) level that fish can safely tolerate depends upon temperature and to a certain extent the species. Volubility of oxygen increases as temperature decreases. In ponds, DO can change dramatically over a *24-hour* period. During the day oxygen is produced by photosynthesis, the process by which green plants convert water and carbon dioxide in the presence of light, to oxygen and carbohydrates. During the night and day oxygen is consumed by respiration, the process by which plants and animals use oxygen to produce carbon dioxide as they burn carbohydrates, but in the day photosynthesis usually produces more oxygen than is used.

Typically, oxygen levels are lowest just before dawn and highest in the late afternoon. DO in a culture system must be maintained above levels considered stressful to fish. Warm water fish (species that grow best at temperatures above 80°F) can tolerate lower DO concentrations than coldwater fish (species that grow best at temperatures below 60°F). As a rule of thumb, DO should be maintained above 3.0 ppm (parts per million; frequently used interchangeably with milligrams per liter, mg/L) and 5.0 ppm for warm and coldwater fish, respectively. Prolonged exposure to low, non lethal levels of DO constitutes a chronic stress and will cause fish to stop feeding, reduce their ability to convert ingested food into fish flesh, and make them more susceptible to disease. Intensive fish production in ponds, cages, flow-through, and recirculation systems requires aeration or oxygenation to maintain DO at safe levels.

Dissolve oxygen concentration is the most important parameter in aquaculture. It is essential to keep organisms living, to sustain species reproduction, and for the development of populations. If the sufficient level of the dissolved oxygen is not maintained, the fish will be stressed and vulnerable to diseases and parasites out break or they will die.

Oxygen is soluble in water inversely proportional to the temperature of the water. When temperature of the water gets lower, the dissolved oxygen level will be the higher. It enters the water by absorption directly from the atmosphere or by plant photosynthesis. It is used out by respiration of organism and by organic decomposition. During respiration and decomposition, animals and plants consume dissolved oxygen and release carbon dioxide. Long period of low dissolved oxygen in pond water can indirectly affect fish by decreasing their resistant to the disease, inhibiting growth. Waters rich in nutrients also produce huge amount algae, which upon decomposition deplete the oxygen supply. Fish kills are often associated with the process of eutrophication.

Objectives

To do and get experience about analysis of dissolved oxygen at the end of the practice.

1. Learn to estimate dissolved oxygen value of the pond water
2. To estimate the total percentage of dissolved oxygen saturation of the pond water.

Protocol

1) Prepare sample

1. Collect water sample, and close the bottle tightly before taking out from the pool, avoid trapping the air bubble in the bottle.
2. Open the sample bottle lid and add 3 drops each of solution A and B. Close the lid tightly and shake well by inverting manner for 10 times.
3. Leave the bottle for complete precipitation for 1 minute shaking again and leave for the precipitate to settle for another 1 min.
4. Add 2 spoon of powder C,close the lid and shake well to dissolve to precipitate.the solution will turn yellow. Avoid this solution into contact with your skin.

2) Titration

1. Add 10 ml of solution into reaction vial.
2. Add 2 drops of solution D and shake well,the solution is turned to violet.
3. Add solution E carefully drop by drop while gently shaking. Stop when solution is colorless.
4. Number of drops used divide by 2 is the value of dissolve oxygen in ppm.

Fig. 16. Test kit used for estimating dissolved oxygen

Field Visits on Small Scale Aquaculture Systems and Fish Feed Mill

10

Aquatic Resources Research Institute (ARRI) Chulalongkorn University

The Aquatic Resources Research Institute (ARRI) was established to conduct research & answer important impending questions associated with the nation's aquatic resources, like conservation of marine organisms, marine science, oceanography, aquaculture, biotechnology etc. It also works for transferring or sharing the knowledge to public, for guiding towards sustainable development.

Fig. 17. ARRI unit

Sichang Marine Science Research & Training Station: (SMaRT) Chonburi Province Thailand

This is another unit of the ARRI. It was established in 1979, in Chonburi Province Thailand. It has gained importance as a role of aquatic resources in national economic development and solving the problems of degradation and conservation of natural resources. Teaching process is carried out for both the undergraduate & graduate l.

Introduction

They have total area of about 35.84 hectare (224 rai in Thai, 1 hectare = 6.25 rai). Research station for marine science and relative researches as well as academic services for local areas. Facilities for researchers like- Laboratory and cultivation areas & have a research vessel named as "Chula vijai 1" Instruments for basic oceanography researches and aquacultures available.

Research Scopes under ARRI includes, Marine science, coastal ecology & oceanography, Marine environment, water quality & pollutions, Conservation for aquatic resources, Research on local fisheries, Application of molecular techniques in marine and relative research, etc.

Fig. 18. Activities at ARRI

Training Station

It is a training station for high school students as well as Undergraduate students in marine sciences both at Bangkok & Sichang units. They conduct programs like- Training students from various universities in Thailand: 20-24 persons/year, Field study areas for various subjects from Chulalongkorn University, Marine science camping (high school student): 1-2 times/year, and Other academic services.

Fig. 19. Sichang training station

Other activities

ARRI also carries out some other activities in the Sichang island facility like-Skin-diving and swimming around Sichang island, Coastal ecology learning: sandy beach and rocky shore, etc.

Mechanical equipment available at ARRI

Laminar air flow

Laminar flow, is commonly used in major laboratories where a sensitive microbial procedure is to be carried out, where only clean air is required. Laminar airflow units create particle-free working environments by sucking air through a filtration system and exhausting it across a work surface in a laminar air stream. They provide an excellent clean air environment for a number of cleanroom requirements.

Inverted Microscope

In an inverted microscope, the source for transmitted light & and the condenser, both are placed on the top of the stage, pointing down towards the stage. The objectives are located below the stage pointing up. The cells here are observed through the bottom of the culture vessel. To meet the criteria for successful inverted microscopy, the bottom of the culture vessel must have optical features. In short, it is a very popular technique for cell imaging. Here live cells are observed through the bottom of the culture vessel. It has several advantages over the upright microscopy.

RTPCR

It is a molecular technique, commonly called as Reverse-transcription Polymerase Chain Reaction. As a starting template, instead of DNA, m-RNA is used. RT-PCR is a modified version of PCR. It's a qualitative means of detecting a particular RNA sequence with a high degree of accuracy and sensitivity.

GC-gas chromatography

It is commonly used in the laboratories. It works on the principle that, the sample solution injected into the instrument enters a gas stream which transports the sample into a separation tube known as a column. Generally, helium or nitrogen is used as the carrier gas.

Gel Documentation

It is a documentation system, used in laboratories, also known as 'gel docs' or 'gel imagers' are used basically to record and analyze the results of gel electrophoresis and membrane blotting experiments. Such instruments are necessary for visualizing stained or labeled nucleic acids and proteins in media such as agarose, acrylamide, or cellulose.

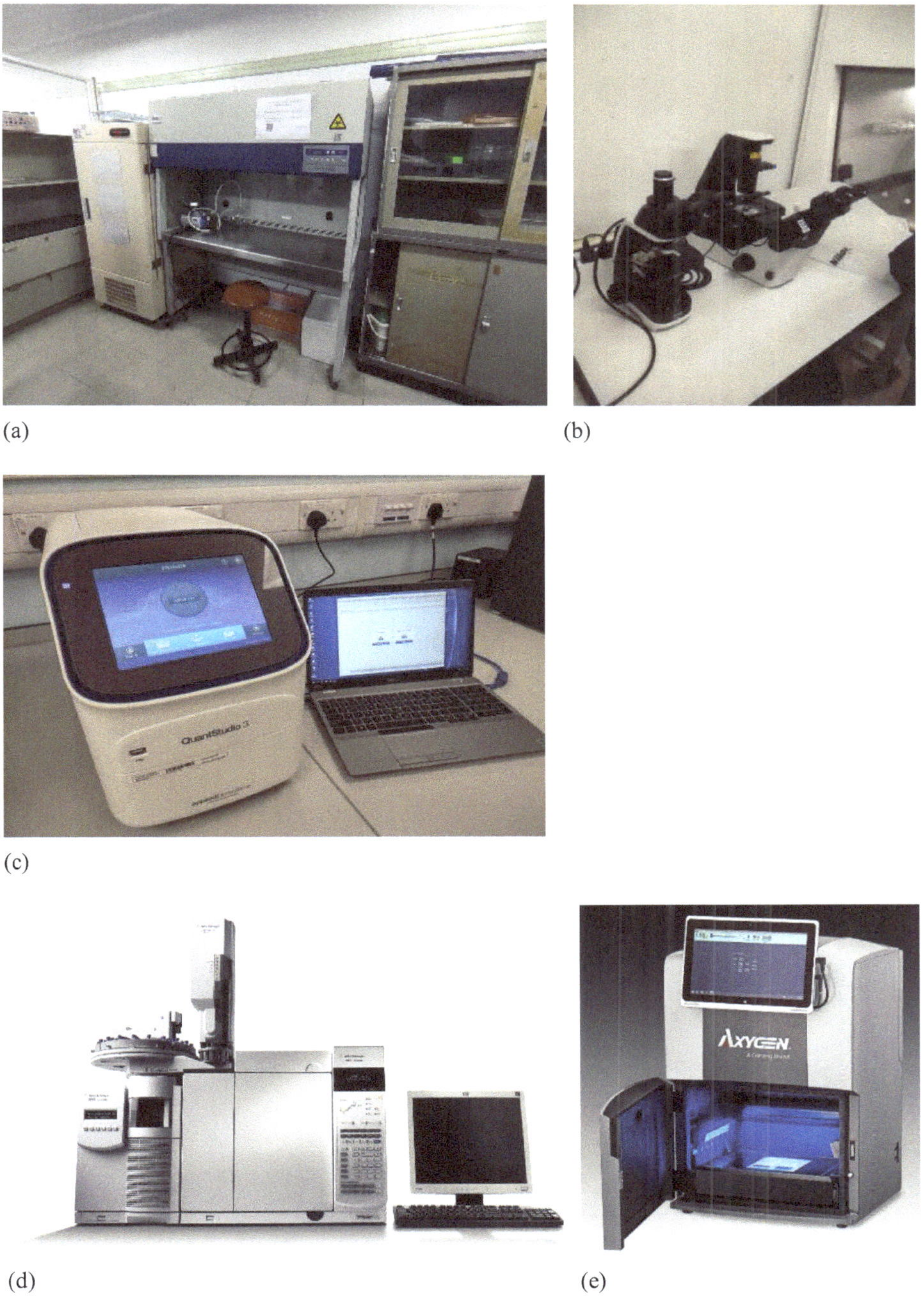

Fig. 20. **(a)** Laminar air flow **(b)** Inverted microscope **(c)** RT-PCR **(d)** GC-Gas chromatography **(e)** Gel documentation

11

Department of Fisheries (DOF), Thailand

Overview of Inland Fisheries Status

In Thailand, inland capture fisheries are carried out mainly in rivers, lakes, swamps and reservoirs. These fisheries are an important source of animal protein for the rural people. Since 2005, production from inland capture fisheries has increased by almost 13 percent. Gillnets, longlines, hook-and-line, scoop nets, cast nets, and lift nets are some of the fishing gears being used. Among these fishing gears, gillnets are the most popular and efficient, particularly in swamps and reservoirs. Thai silver barb, snakehead, walking catfish, local carps, and Nile tilapia are caught. Nile tilapia, hybrid catfish, silver barb, snakeskin gourami, giant river prawn, striped catfish and striped snake-head contributed 95 percent in quantity and 92 percent in value in fresh water aquaculture. But due to urbanization and industrialization of the countryside, impacts on natural resources, including water resources. The size of freshwater fish stocks depends largely on natural circumstances.

Reference collection room of freshwater fisheries

It is a taxonomic unit lab with catalogue number 3000. It consists of freshwater fish specimens as well as few brackish water fish specimens. The collection contains variety of fishes including string rays, bony fishes etc. The specimens are obtained from the research vessel or from the local fish farmers. Later stored in 10% formalin and 30% ethanol solution.

Fig. 21. Visit to Aquatic Animal Health Research Development Department (AAHRD)

12

Model Sea Farm: Phetchaburi Coastal Aquaculture Research and Development Center

We visited The Model Sea Farm Project under the Royal Initiative of Her Majesty Queen Sirikit on 10/10/2023 at 1 pm It is a government farm located at Phetchaburi Coastal Aquaculture Research and Development Center. The purpose of the visit was to learn more about seaweed production and zero waste aquaculture system.

1) Sea ranching

Species used for sea ranching was Banana shrimp (*Fenneropenaeus merguiensis*). 2 months old shrimps were collected from tank and used for ranching.

2) Broodstock maintenance in cage

Cage was having depth 1m and diameter 6m. Cost of production of cage is 1 lakh baht. Stocking density of both cages are 50 no.s. 4 years old brooders of milkfish (*Chanos chanos*) was kept in cage 1. Each fish have 1.5 kg weight . 6 years old brooders of Silver pompano (*Trachinotus blochii*) was kept in cage 2. Each fish have 2 kg weight. Feeding done only one time during morning hours. Commercial feed containg 42% protein content was used for feeding. No other maintenance was done to cages. During breeding, brooders are transferred to another tank.

3) Raft culture of Green mussel (*Perna viridis*)

Raft details

Raft used is movable and floating. During rainy season salinity of the coastal areas get reduced, so farmers move the rafts in deeper areas. Area of raft is 20 m^2 .300 no.s of ropes were used for making raft. Rope material was HDPE, nylon. Rope dimension was 24 mm,16mm and 6mm for outer rope , for inner rope and for hanging rope respectively. 300 No.s of barrels were used as floats.

Concrete sinkers having 2 tonnne weight were used as sinkers.Average depth of rope was maintained as 1m. Raft also have extra modification of fish trap at the middle. Fishes are also collected from the trap during harvesting of mussels.

Culture details

Seeds were collected from wild. During 8 months culture period, growth rate varies according to season. Sampling was carried out once in a month. 200 kg partial harvesting is usually done during sampling. Estimated yield of 2 tonnes per crop per year is obtained. Good quality mussel after removing barnacles have selling price of 20 baht /Kg. Moderate quality mussel is sold at 10 baht / kg. Conventional method of pole culture was also practiced.

4) Zero waste aquaculture system

Seawater was pumped from the sea as per need. Pump have the capacity of 3hp. Whole area is divided into 5 zones of different salinities and ponds in each zone are utilized for different puposes. Parameters like nitrite, nitrate, ammonia, alkalinity and pH are measured 2 times in a week. Floating feed were fed 2 times at 10 AM and 4PM. At 5% of BW. Herbivorous fish feed contains 18% protein content whereas carnivorous fish feed has 42%. No any fertilization is done for ponds.

Zone 1 consist of 3 ponds for Asian seabass, milkfish and sea grapes respectively. The salinity of ponds is maintained at 25-40 ppt.

Zone 2 also maintained at 25- 40 ppt and two ponds are used for culturing seagrapes and one for seabass.

Zone 3 have salinity 80 to 120 ppt. It is used for Artemia culture. Paddle wheel aerators are used for 3 hours from 9 to 12 AM. Mono sodium glutamate and molasses added to increase plankton growth.Partial harvesting is done during morning hours by using scoop net.

Zone 4&5 maintained at salinity 5 – 25 ppt. This area is used for grown out culture of euryhaline fishes. These are restricted areas.

Separate area is used as salt pans. After complete evaporation, salts and many other products are prepared.

5) Seaweed production unit

Cultured species: *Ulva rigida*(Sea lettuce), *Caulerpa lentillifera* (Sea grapes)

Culture method: It is cultured in tanks having 500 lit. capacity, having salinity 30ppt, ammonia less than 0.4 ppm, alkalinity in range of 120 to 150ppm. 0.5 Kg of seaweed is spread in a hanging net. Culturing it for 2 months gives growth of

10 kg. After harvesting, seaweeds are kept in freshwater for 8 hours with aeration. Later for 24 hours in another tank. Now it is ready to eat. It is sold in the market at the rate of 150 baht / Kg.

Uses: These are having wide usage in various industries like, Ice creams, protein bar, cosmetics, soaps etc.

6) Products prepared from the farm

Raw salt (having sweet and salty taste) is prepared from salt pan. $MnSo_4$(Epsom salt), Table salt (having only salty taste.), Gypsum, dried algae(anything that accumulate under the pond) used as manure, natural seawater powder which is a mix of calcium carbonate ,magnesium and all other minerals in sea (150 baht /Kg), scented salt body scrub, natural salt face powder aromatic soap are produced.

Fig. 22. Various activities at Model Sea farm

13

Tilapia Floating Cage Culture Sukjai Plakarn Farm

Construction details

They had a total of 36 tanks, and most of them were made of tin. Each tank had a diameter of 6 meters, was 1.2 meters deep, and could hold up to 30 tons of water. Inside these tanks, there was a plastic lining. Additionally, there were 4 to 5 tanks made from bamboo. Bamboo tanks have a shorter lifespan, lasting 1 to 2 years, while tin tanks can last for 8 to 10 years.

Culture details

The species that was cultured over there was Red tilapia. Culture practice was intensive. The nursery-rearing period was 70 days. In the first step, fish were raised from 0.5 grams to 10 grams in a separate tank for 35 days. The second step involved the growth of fish from 10 grams to 30 grams for the next 35 days in another separate tank. The fishes were bought from outer sources and they were all male fishes. The temperature of the water was 23° C, pH was 7 and the dissolved oxygen was 7ppm. For filtration they used to lower the water level, then the water pump is turned on. This made the water flow in circles. The dirt and waste are gathered in the middle of the tank. There was a special outlet in the middle to remove the waste. After that, the dirty water is sent to a treatment tank, where it is cleaned with chlorine. Finally, the cleaned water is released back into nature. Aeration was achieved using pipes installed at the bottom of the tank. For small fish (<10 grams), an 80% water exchange was conducted twice daily. For larger fish (>10 grams), a 200% water exchange was carried out once a day. Common diseases which they have observed are Trichodina and Streptococcus. After 70 days of nursery rearing the fishes of 30 grams are released into grow-out cages.

They also had one processing unit. In the processing units, they have a mixer, sausage machine, cutter, fish ball machine, oven and vacuum machine. One of their notable products was Tilapia fish sausages. They primarily sold fish and fish-related products through online channels. The construction cost of that processing unit was around 1,80,00,000 INR.

The purchase price for each fish was 2.7 Indian Rupees. The installation cost for a single tank made of tin was 93,000 Indian Rupees, while for a bamboo tank, it was 34,000 Indian Rupees. They procured 400,000 fish seeds every month.

Fig. 23. Activities at Sukjai Plakarn Farm

14

Tilapia Grow-out in Earthen Pond Kanjanaburi Province

The ITAF participants visited the Anuban Tilapia Floating Cage Culture on the Canoy River situated in the central part of Thailand on 11/10/2023 at 1:30 p.m. The purpose of the visit was to learn about the floating cage culture on a large scale.

Construction Details

The farmer had a total 500 number of cages divided into 5 units i.e. 1 unit contains 100 cages. The dimensions of the floating cage are 7m in length, 5m in width and 2m in depth having an area of 70 m3. The fabrication of the cage was done by using the GI pipes attached to the barrels for buoyancy. The double HDPE netting was done from five sides and on the upper side bird net was fixed as a biosecurity measure. The construction cost of the cage was 140000 baht /cage.

Fig. 24. Floating cages

Culture Details

The all-male red Tilapia were cultured on this farm. The fingerling of 0.35g is stocked in the nursery cage and grown up to 20g. It required 2 months, in the 1st month grew up to a size of 5g having a survival rate 70% then 50% of these were shifted to another cage. From 5g to 20g, it requires the next 1 month having a survival rate of 50%. The 20g Tilapia had shifted in the grow-out pond having a stocking density of 35 nos./m^3. During the 6 months, it had grown up to the marketable size of 1 Kg having 80-90% survival rate in the grow-out floating cages. The total production from a cage was 2.5 tonnes/cage. The price for 1Kg fish in the market was 80 baht and sometimes goes upto the 90-100 baht maximum. For the daily feeding and maintenance, 2 labors were working but this farmer had 500 labors on rotation basis in 5 units. A lot of 150 tonne fish were harvested every month and restocking of 2500 fingerlings was done batchwise. The FCR was 1.8 which was resulting into the 6g average daily gain. Feeding was done 3 times in a day with complete commercial feed containing. >30% protein for all stages of fish. The net profit from each cage was 40000 baht and 5M baht was annual profit. For the parasite treatment $KMnO_4$ solution was used. The major constraints faced by the farmers were the 1-2 months risk of flood. To escape from this, the cages were drawn towards the bank side to avoid the fast-flowing water and cement bags were used as the anchors to hold the cages steadily. As per the rule, farmers have to pay 500 baht/cage/year to the Military Welfare Benefit as the lease. The farmer had suggested doing the nursery rearing and grow-out culture both in the river to avoid excess mortality.

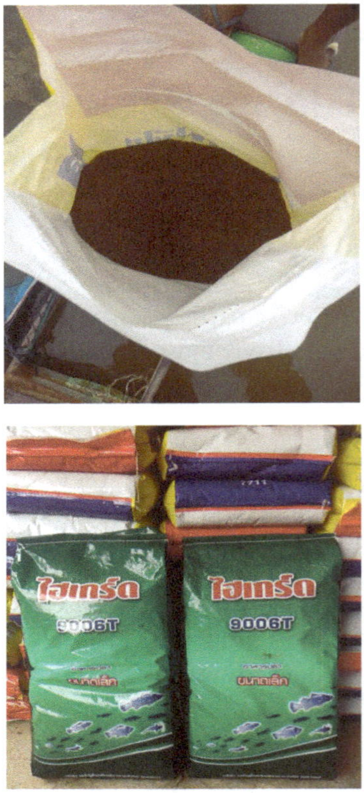

Fig. 24. Commercial feed

Fig. 25. Discussion with the farm owner

15

Feed Mill Feed Formulation and Feed Production: CPF

We visited the CPF (Thailand Public Company limited) in Samutsakorn on 12/10/23 at 1:30 pm. The purpose of the visit was to know about the feed ingredients and production process.

1) Overview

Product like shrimp feed and fish feed are prepared by the company. Company was spread in 61 Rai(24 acre) area. 400 employees are working in the company.Working hours is 3 shift / day and working days – 6 days / week.Shrimp feed unit have the capacity of 13,000 T/M and actual production was 7000 T/M. 8 line pellet machine was used for this purpose. Fish and frog feed unit have capacity of 11,000 T/M and actual production 7000 T/M. 6 line extruder machine was used.

2) Production flow

Raw materials are collected in truck scale and 100% sampling is done. If it pass the quality check it is stored for further procedures or else it is rejected. After intake of raw materials , coarse grinding is done and weighed. Later first mixing is done. After fine grinding second mixing is done.

Pelletization process proceeds by steaming,drying ,cooling, sizing and packaging. Extrusion process proceeds by drying, coating, cooling, sizing and packaging.

3) Experiments done by company

Presently experiments on vegetarian feed development activities are carried out.no fish meal and fish oil is used in this diet. Also no animal protein and oil. Gives similar or better growth performance as that of animal feed diet. Both are have same FCR.

Attractant trials for developed feed are also performed by the company. For this in 30 l tank, 20 pieces of shrimp having weight 2.3 – 3 g are kept. 5 minutes testing time is followed 3 times a day. 95% water exchange is done. 3 replicates of attractants are used and shrimps are counted within specific time period.

4) Factory visit

Capacity of warehouse is 3000 tonnes .Types of feed – floating fish feed, sinking feed for shrimp and also of feed according to different life stages and species. Shrimp feed have the bulk density of 600 g/l whereas fish feed have 400g/ l.2categories of feed- premium feed for specific species ,economy feed- for multiple species are manufactured. The product flow from raw materials to final end product is controlled from fully computerized control room. Robotic system are used for lifting the bags. The Water boiler unit produce the steam and circulate to production unit. Raw materials are collected from countries like Brazil, Argentina after a series of quality check inspections. Only Tuna fish is used for fish meal as per international guidelines. A technical service team is set up for the follow up activities like assistance to farmers and conducting customer satisfaction surveys every 3 months. Feed is mainly exported to countries like India, Vietnam . Production of 20000 tonnes/ month from one plant and 500000 tonnes / year from all plants are attained.

16

Snake-skin Gourami Sukanya Farm

On the morning of 13 Oct 2023 morning (9:30 am - 12:00 pm), ITAF participants visited snake-skin gourami local farms in Samut scorn province. Snake-skin gourami is one of the best delicacies in Thailand which has large demand in local and international markets. Snake-skin gourami can breed after reaching 100-200 gms (2 months). The culture period of snake- skin gourami is 9-10 months. In snake- skin gourami males have a large body and long dorsal fin than females.

Fig. 26. Sexual dimorphism

Fig. 27. Transportation route through canal

Farm 1

Snake-skin gourami culture is done in 30 rai (1 rai= 1600 m2) area of the farm. The depth of the pond is 1.5m. A single crop is taken in one year. 50,000 no. of fry are stocked in 30 rai area. Feeding areas are constructed by the net embankment where feed is given to fish harvesting is done after 8-9 months. The cost of production for one crop is 2 million/year in that they produce 30-40 tonnes /30 rai, which is almost 3 million / year.

Fig. 28. Culture pond of snake-skin gourami with feeding area

Farm 2

This farm culture broodstock of snake-skin gourami. it is a breeding as well as a nursery pond. depth of the breeding ground is 50 cm. The broodstock ratio in the pond is 3 males: 1 male. 3 weeks after breeding, the seed is collected. Natural floating creeps work as a substrate for eggs. for seed collection, they are using a seed harvesting machine.

Fig. 29. Seed harvesting machine

Fig. 30. Breeding as well as nursery pond

Farm 3

200 broodstock fishes are stocked in the breeding pond. the area of the pond is around 5 ha.

Fig. 31. Nursery pond of snake-skin gourami

17

Small-scale Frog Farm

Frog farming in Thailand started 20 years ago. The frogs commonly found and farmed in Thailand are the local species, *Rana tigerina* and *Rana rugulosa*. The first imported bullfrog, *Rana catesbeiana* was introduced in 1980 and is popular in the northern part of the country. *Rana tigerina* is olive brown in colour which is about 15cm long when fully grown. The adult weighs about 300g. *Rana rugulosa* is another local species that is olive-brown in colour with a dark head. It is about 13cm long when fully grown and weighs about 270g. The imported species, *Rana catesbeiana* is olive-brown to green in colour. It is 17 cm long when fully grown and weighs around 459g.

The farm contains a total of 4 earthen ponds of 1 Rai (1600 m2) each. The stocking density is 100,000 tadpoles/pond. The tadpoles are reared for 6 months until harvest. The water quality of the pond is maintained by the daily cleaning and addition of water to maintain the water level. The tadpoles are fed with food pellets of 40-45% protein content. Feeding is done at the rate of 20% of body weight while in adult frogs of around 1 month old, feeding is done at the rate of 4-5% of body weight. Each frog weighs about 150-200g at the time of harvest. The frogs are harvested in live condition and packed in 10kg bags to sell in the local market. The process of post-harvesting is initiated by the removal of sludge manually. Followed by the application of lime in order to increase alkalinity and destroy the bacteria present in the pond. The entire process of post-harvesting takes around 20 days, to begin with the next cycle.

The seed costs around 0.1 baht/tadpole. During breeding season, the cost of seed is around 0.05 baht/tadpole. The production cost per kg is around 48-50 baht. During low season the frogs are sold at the rate of 55 baht/kg but during high season it goes around 100 baht/kg.

Fig. 32. Earthen ponds for frog farming

18

Sriracha Fisheries Research Station Faculty of Fishery, Kasetsart University (Sriracha)

We, the participants of ITAF2023 visited Sriracha fisheries research station under the guidance of Asst. Prof. Dr. Naphat Panthukumbol and Dr. Somrat Taweedet on 16/10/23 at 08:00am. The research station is under faculty of fisheries, Kasetsart university in Chonburi Province. The purpose of visit was to learn about the mussel aquaculture and Integrated Multi- Trophic Aquaculture (IMTA).

Fig. 33. Participants visiting raft culture area

1) Mussel aquaculture

Cultured species was *Perna viridis*(Green mussel). Almost all rafts were owned by fishermen and 10 rafts were under the management of sriracha fisheries research station. The cost of production was 100 baht /acre. Simple method of culture is practiced by installing rafts before breeding season and allowing the natural spawn to attach. The harvesting size of mussel is 8 cm. Observed growth rate is 1cm/ month. Other maintenance includes repairing of ropes ,buoys etc. during off-season. Farmers face many constraints like plankton bloom, waste water discharge, predation by sting ray,parasitism during summer months. Production reported is 10-20 tonnes/raft / yr. Mussels are sold directly by fishermen at the selling price – 25-40 baht. In market ,after cleaning it is sold at 60-70 baht. 200 – 500 farmers manage the culture. For harvesting rafts under research station, 5 baht / Kg is given as harvesting wage to fishermen.

2) Raft details

Total 1000 rafts are installed in 300 acre area. Rectangular shaped rafts are used for culture. 8 anchors weighing 1 tonne each is used for one raft. 4 are arranged in inner corner and 4 in outer corner. 1.2 m distance is maintained between adjascent buoys. Depth of rope is changed according to water depth. Used fishermen net is used for preparing ropes. This will reduce the installation cost. No.of ropes arranged in 1m^2 is 5. Total area covered is 200 m^2. No. of ropes in 200 m^2 is 1000. Buoys are made using HDPE barrels. Approximate shelf life of raft is 5 years.

Small boat is used for sampling. Nearby island protect the culture system from strong waves and act as embankments. In female mussel orange color eggs are present whereas male contain white color sperm.

3) Monitoring System

The upper part measures precipitation, wind direction, wind speed, temperature, light intensity and lower part measures turbidity, pH , salinity.Acoustic Doppler current profiler (ADCP) measures speed and direction of currents.

4) Integrated Multi Trophic Aquaculture (IMTA)

Solar energy is used for working of aerator and autofeeder. 20 m distance is maintained between cage and raft. Seabass or grouper kept in cages.Bottom side is used for culturing sea cucumber and sea urchin. Rafts at the side are used for growing seaweed and green mussel.

Fig. 34. Activities at Sriracha fisheries research station, Kasetsart University (Sriracha)

19

Institute of Marine Science Burapha University, Bangsean Province

On 17 October we visited the Institute of Marine Science, Burapha University (BIMS), THAILAND at 9 a.m. The institute is situated in Bangsean. BIMS is a marine research institute of Burapha University. They have 6 organizations: Marine Research Department, Marine Aquarium Department, Marine Museum Department, Academic Service Department, Office of the Director and Research Station.

They have shown us their Life Support System of Public Aquarium and coral propagation and conservation.

1) Public aquarium

We have seen the Life Support System of Public Aquarium i.e. Filtration System, which involves the complex systems and technologies used to maintain the health and well-being of marine life in public aquariums. This is crucial for keeping the animals and ecosystems on display healthy and sustainable. The display tank has having 10,00,000-liter capacity. Maximum stocking biomass was 5 kg/m3 (3-5 kg/m3). Cartilaginous fishes(Sharks and Rays) and Bony fishes(Trevally and Grouper) are kept in the display tank.

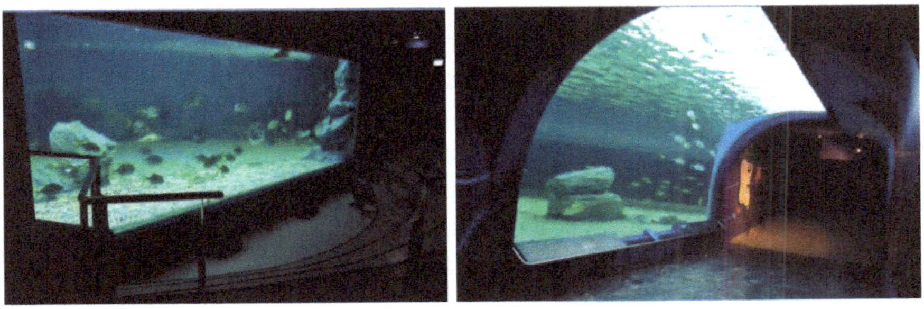

Fig. 35. Display Tank

Fig. 36. Species kept in the Aquarium

Table 1. Environmental requirement of the species

Salinity	32-34 ppt
Temperature	28-30 °C
pH	8.0-8.4
Dissolved oxygen	≥ 5 ppm
Alkalinity	80-120 mg as CaCO3 /L
Total ammonia nitrogen	< 0.1 ppm
Nitrite nitrogen	< 0.05 ppm
Nitrate-nitrogen	< 100 ppm

They were using The Second Generation Filtering System which is designed by BIMS itself. Very efficient use of space was there and filters occupied only a small area. The circulation rate was 15 cycles of display aquarium volume per day (15,000 m3/ day). Removal of solids and suspended solids has been done in the Sedimentation tank(primary sedimentation)and Sand filters. Dimensions of the sedimentation tank were 5m x 7m x 2.5m with 42 m3 Inclined Plate Settlers (Tube Settlers) with 630 m2 sedimentation area. The water depth in the tank was 2m. We've also fed fish in an aquarium. They gave each of us a can with 6 to 7 small sardine fishes for feeding to fishes.

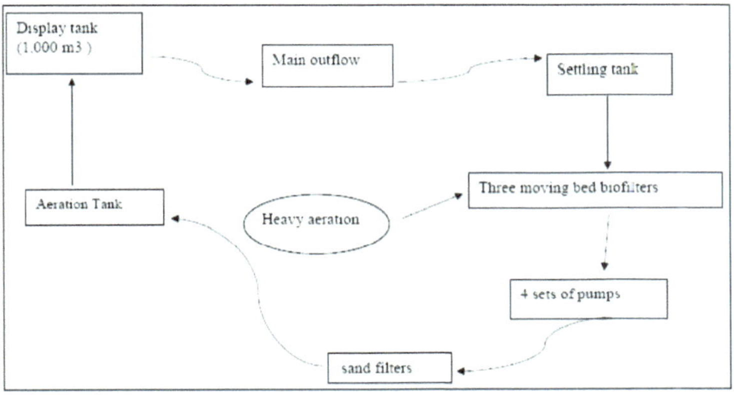

Fig. 37. Schematic representation of the generation filtering system

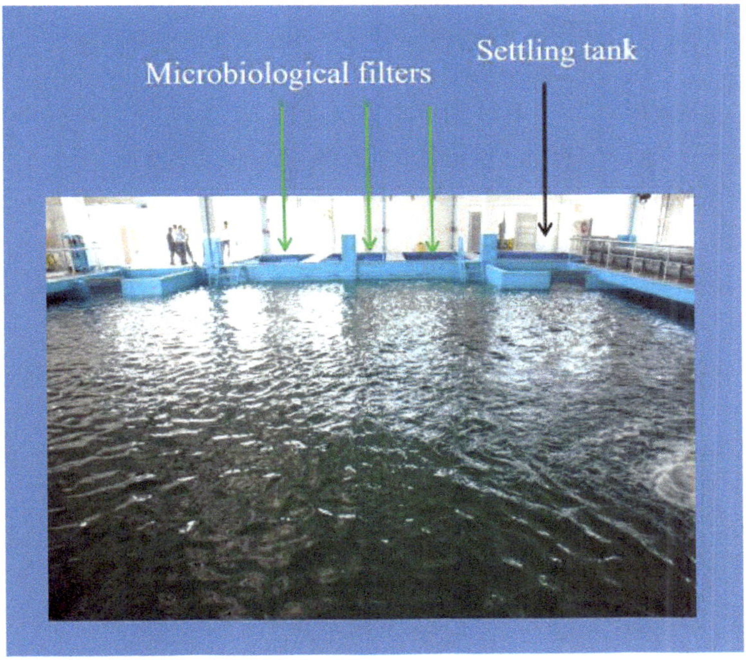

Fig. 38. Display, settling and microbiological filter tank

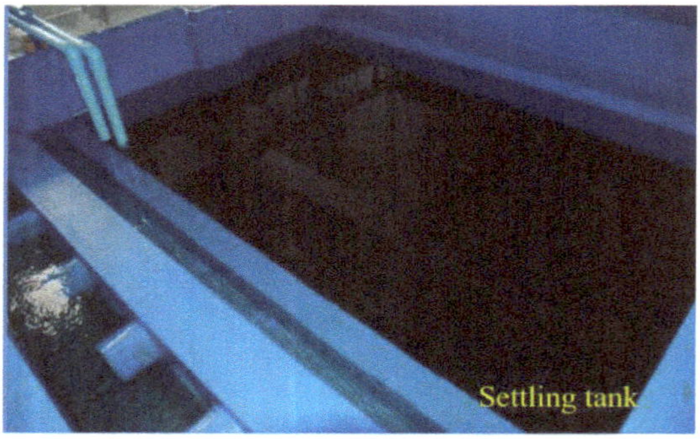

Fig. 39. Settling tank

Fig. 40. Moving bed biofilters

Fig. 41. Sand filter

Fig. 42. Pumps

Fig. 43. Feeding aquarium fishes

2) Coral Propagation and Conservation:

We have seen coral reef propagation and conservation in the BIMS Marine Research Department. Research on Coral propagation is currently going on in BIMS. 30 to 35 HDPE tanks were there. Many species of Hard corals and Soft corals were kept in the tanks. Continuous aeration is provided to the tanks. The bubble that has been generated due to the aeration was big so that the water should not remain stagnant. The temperature in the water tanks was 26°C and the pH was neutral. Coral reproduction had been reproducing naturally in their environment.

They were using the Recirculatory Aquaculture System(RAS) to keep the corals healthy and in good condition. RAS consists of the Plain settling tank, Seaweed biofilter tanks and $CaCO_3$ enrichment tank. By combining all of these elements, the BIMS Marine Research Department is working to create an optimal environment for coral propagation, growth, and conservation.

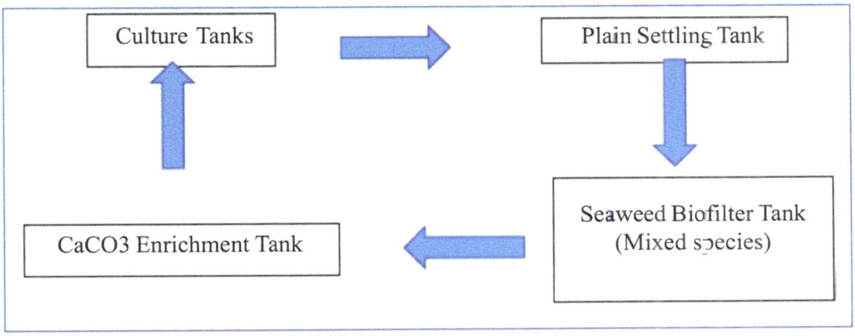

Fig. 44. Schematic representation of

Fig. 45. RAS for Coral Culture

The capacity of the Plain Settling Tank was 150-300 m3/day. The settling area was 3.75m2. In seaweed biofilter tanks, seaweeds are used to export nitrogen and phosphorus out of the system and it prevents nitrate accumulation from nitrification. This is designed for a low-cost easy-operation system. The dimensions of the tank were 6m x 2m x 1.5m. *Caulerpa lentillifera* and *Caulerpa serrulate* are the species of seaweed kept in tanks. The area of seaweed biofilter was $12m^2$.

Daily ammonia nitrogen excreted was 135 g NH3 -N.

Fig. 46. Variety of coral species in Coral culture tanks

Fig. 47. Coral Culture tank

Fig. 48. Seaweed Biofilter Tank

Fig. 49. ITAF Team at BIMS

20

The Ocean Farm Shrimp Nursery

We, the participants of ITAF2023 visited The ocean farm under the guidance of Asst.Prof.Dr.Naphat Panthukumbol and Dr.Somrat Taweedet on 17/10/23 at 3:00 pm. The farm is certified by fisheries department and is located in Chonburi Province. The purpose of visit was to learn about the nursery rearing of shrimp and use of frozen artemia.

Fig. 50. Participants with ocean farm team

1) Nursery rearing details

The cultured species was *Litopenaeus vannamei*. Rearing was done from nauplii to PL15 stage. The farm is producing 2 variants named- 'The ocean hero' which is disease resistant and 'The ocean speedy' having high growth rate. They are procuring nauplii from Phuket. Then double certification from DOF and Burapha university lab is done to ensure SPF stock. The purchasing price of nauplii is 12,000 baht/ million nauplii and transportation done by plane. The culture period of shrimp is around 21 days.Tank used for rearing have the dimension 5 × 1 ×1 m. Chaetoceros was fed during initial days and later artemia was used. A feeding frequency of 6 times a day is mainatained.Parameters like salinity, alkalinity,nitrite,ammonia,pH,O2,turbidity is measured daily and proper records are kept. Addition of brown color in rearing tank water to prevent cannibalism is also followed. Survival rate of 100% ensured by all these practices. The farm have production capacity of 18 million PL.1.26 million baht/ batch production cost is estimated. 0.12 baht / nauplii is the selling price. Selling of PL is carried out in upper eastern,central &southern regions of Thailand. 20-30% profit is generated per batch.After each cycle of rearing, an interval of 15 days is given for cleaning tanks.

Fig. 51. Reared *Litopenaeus vannamei*

Fig. 52. Brown color mixed in rearing tank water

2) Chaetoceros culture tank

The salinity is maintained at 30 ppt. For culture tank 50% SW mixed with 50% FW. The source of chaetoceros is procured from Burapha university and are inoculated into the culture tanks.

Fig. 53. Chaetoceros culture tank

3) Frozen artemia

The farm is procuring frozen artemia box from I&V BIO INSTART company. It is stored at 4°C. The shelf life of the product is 2 days. As per need , product is mixed in small quantity of water and it is ready to feed.

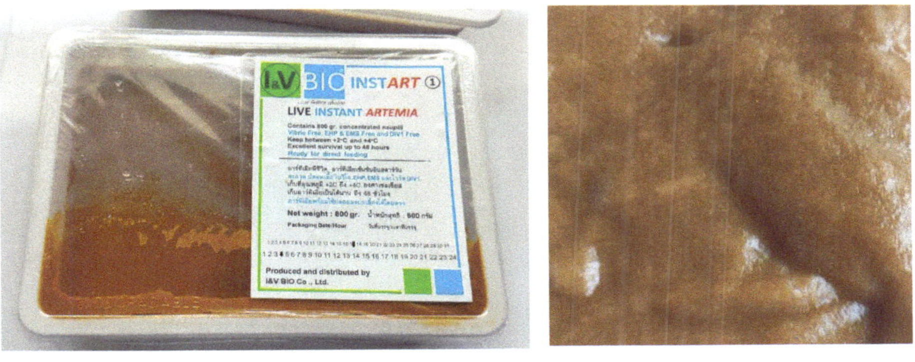

Fig. 54. Frozen artemia box

Fig. 55. Frozen artemia mixed in water, ready to use

21

Boonsawang Seabass Farm Seabass Brood Stock and Hatchery Management

Introduction

We, the participants of ITAF2023 with Dr. B. R. Chavan and Dr. Prashant Bodake, visited Boonsawang Co. under the guidance of Asst. Prof. Dr. Naphat Panthukumbol and Mr. Suthi Mahalao on 18/10/2023 at 09:00 am. Boonsawang Co. in Chonburi, owned by Suthi Mahalao, also exports fry to Vietnam, Malaysia, Singapore, Taiwan, Australia, and Israel. In addition, Suthi runs his sea bass farm and is a feed distributor to farms in the province. The purpose of the visit was to learn about the seabass culture system and get familiar with technical aspects. The farm occupies an area of 10 hectares and is covered in 11 ponds. The farm is aimed to grow from fry to marketable size seabass. The single crop duration is divided into 4 periods.

Period 1: Fry to Fingerlings (1.5 inch)

During Period 1, spanning 40 days, the focus at the Fry Nursery Farm was on the crucial transition from fry to fingerlings. This critical stage involved specialized training in feeding young fish, ensuring their growth and development. The farm's dedicated team of experts closely monitored and nurtured the delicate fry population, aiming for optimal growth and health. Fry size sorting was another integral part of this process, where careful attention was paid to selecting 1.5-inch fingerlings, ensuring that the fingerlings were of the right size for their future growth and development. Nursery rearing in 5 ppt saline water. Size grading is done in intervals of 4 days. If grown in more salinity, slow growth is observed. This phase is carried out in an indoor system.

Period 2: Fingerling (1.2-1.5 inch) to Fingerling (4-5 inch)

During Period 2, which spans 30 days, the focus is on the transition of fingerling fish from the initial size of 1.2-1.5 inches to a more substantial size of 4-5 inches. Fingerlings from an indoor system are stocked in an earthen pond in a small compartment made with a fine mesh net. This gets done to mark the feeding area and reduce feed loss. The Feeding area gets extended according to growth. At the outset, the fingerling fish are about 1.5 inches in size, and they are quarantined in the spot where they were initially placed in the soil pond. This quarantine period helps monitor and ensure the well-being of the fingerlings. Feeding the baby snapper is crucial during this period, as their nutritional needs evolve as they grow.

To effectively manage the growth of the fish, drag screening is employed to separate the fingerlings that have reached the size of 3-5 inches. The larger fingerlings are then selected for further growth, while the smaller ones are left to continue developing. The size of the basin used for this screening is carefully chosen to accommodate the selected sea bass. Additionally, it is essential to keep track of the number of fingerlings at this stage, as it aids in the management and delivery process to the farmers.

Fig. 56. Quarantine at the spot where the baby fish is placed 1.5 inches in the soil pond.

Fig. 57. Feeding Fingerlings

Period 3: Fingerling (4-5 inches) to Marketable size (600-900 gm)

During Period 3 of fish farming, the transition from fingerling-sized fish to marketable adults is a crucial phase that spans approximately 150 days. Successful preparation of the ponds and the aquatic environment before releasing the 4–5-inch baby fish is paramount for their healthy development. Pond floor treatment with a chain vessel ensures the removal of any debris or contaminants that may adversely affect the water quality. The FCR observed during this period is 1.2.

Fig. 58. Preparation of ponds and water before releasing fish

Before releasing the young fish, meticulous pond preparation is required. The ponds should be adequately aerated, and water parameters such as temperature, pH, and dissolved oxygen levels must be carefully monitored and adjusted as needed to create an optimal habitat for the growing fish.

Husbandry rules are instrumental in ensuring the well-being of the fish during this period. To achieve the best results, it's essential to use 4–5 inch baby fish that have been trained to eat kibble. These specially trained fish are better adapted to the commercial feed provided and are more likely to thrive during their development.

Feeding the fish over the 4–5 month period is a vital aspect of their growth. The Average Meat Exchange Rate (FCR) is maintained at a low 1.2, indicating efficient conversion of feed into fish meat. Feeding schedules typically consist of 1-2 meals a day, depending on the size of the fish. Additionally, feeding is done in low-light conditions to reduce stress on the fish. Feeding times are usually early in the morning from 6.00 to 6.30 hrs and in the evening from 17.30 to 18.30 hrs. These practices help ensure that the fish undergo a healthy and efficient transformation from small fingerlings to market-ready adults, ready for harvest.

Period 4: Table size (900 gm) to Jumbo size (3-5 Kg)

Over one year after the previous period, the transformation from a table-sized seabass, typically weighing around 900 grams, to a jumbo-sized seabass, ranging between 3 to 5 kilograms, is a remarkable growth achieved by this farm. The FCR recorded in this period is 1.4.

Stocking density

For the first 5- 10 days stocking density kept is 1,00,000 fry per m2. For fingerlings, the stocking density is 30,000 per m2. For adult fish, 3 per m2.

Feed

Starting with 10% BW for fry and reducing to 8% in the later life stage.

Survival rate

The overall survival rate informed by the owner is around 80%.

Production

20 tones

Protein requirement

Starting from feed containing 55% protein to fry and reduced to 35% in later stages.

Water quality for farming

Oxygen solubility (DO): greater than 3 mg/lite

pH: 7.5-8.5

Alkalinity: greater than 100 ppm

Salinity: 0-30 ppt

บ่อ	pH	Ammonia	Nitrite	Alk	ความเค็ม
1	8	0	0.1	>170	20
3	8	0	0.1	136	28
4	8.3	0	0	85	22
8	8	0	0.1	>170	27
9	8.3	0	0	102	31
10	8	0	0	136	26
11	8.3	0	0	153	29

Management during farming

Dip treatments with a formalin concentration of 100 ppm for 5 minutes are a common practice to control parasites and bacteria that can cause diseases like gum rot, saltwater spots, and freshwater conditions. This preventive measure helps keep the fish population in good health and ensures that they thrive in their controlled aquatic habitat. In addition to disease prevention, the feeding regimen is another critical component of aquaculture management.

Problem

The problem of black-skinned in seabass can be attributed to several factors that affect the appearance of the fish and market price.

Solutions

The black or dark discoloration of the water which causes this problem, is often a result of various factors, including the accumulation of organic matter, suspended particles, and the growth of algae and shellfish. To address this issue, several solutions can be implemented. Firstly, shellfish removal can be an effective method to reduce dark coloration, as they often contribute to the buildup of organic matter in the water. Secondly, controlling and managing the growth of algae through proper treatment and regulation can significantly improve water quality. Additionally, regular maintenance of water bodies, including cleaning and removing debris, can help maintain water clarity and reduce the black-skinned appearance. Monitoring and adjusting the water level and depth can also play a role in improving water quality and reducing the intensity of the dark color.

Where does the cost of fish farming come from???

The cost of fish farming encompasses a range of factors that contribute to the overall expenses involved in this aquaculture practice. One significant expense is land rental, which typically falls within the range of 1,500 to 2,000 baht per rai per year. Utilizing land for fish farming is a crucial aspect, as it provides the necessary space for ponds and infrastructure. Additionally, fish farming relies on water resources, including natural water sources and ponds, which come with their own maintenance and operational costs.

Energy expenses also play a significant role in the overall cost, encompassing the use of oil and photovoltaic systems for powering various equipment and maintaining suitable environmental conditions. Another essential component is the quality of fish breeding and the source of breeding materials, which can affect the success of the operation. This includes factors like breeding balls' quality and the health and genetics of the fish being cultivated.

Food and bait are essential costs, as they directly impact the growth and health of the fish. Fish farmers may use a combination of fresh bait and dry feed, both of which come at a price. Labor costs, including the employment of human workers for tasks like feeding, monitoring, and maintenance, also contribute to the expenses. Maintaining water quality and conditions in the ponds is crucial, necessitating investments in water quality improvers and water conditioning mortars to ensure the health and well-being of the fish.

Increasing profitability in fish farming

Increasing profitability in fish farming is a crucial objective for fish farmers, and one key aspect to focus on is the Feed Conversion Ratio (FCR). FCR, often represented as a ratio, measures the efficiency of converting feed into fish meat. A lower FCR indicates a more efficient process. For example, an FCR of 1.2 to 1.4 means that it takes 1.2 to 1.4 kilograms of pellet feed to produce 1 kilogram of fish meat. Lowering FCR can significantly reduce feed costs, which is typically the largest expense in fish farming. Additionally, utilizing high-quality pellet feeds can help achieve better FCR values. It's important to note that FCR values can vary depending on factors like fish species, water quality, and farming practices. To further enhance profitability, it's essential to consider feed quality, management techniques, and sustainable practices while keeping in mind that in some cases, using fresh bait fish can also be an efficient option, where 5 kilograms of bait fish are equivalent to 1 kilogram of dry feed, potentially leading to cost savings and improved profitability in fish farming operations.

What does it cost to raise fish to a weight of 1 kilogram?

Type of food	Food price/kilogram	FCR	Cost of food	Fish breeding balls	management	duration (month)	summarize
Dry feed	50	1.2	60	8	12	4.5-5.5	80

In Baht

What damage to farming

In aquaculture, plankton drops can disrupt the delicate balance of aquatic ecosystems, potentially leading to decreased water quality and reduced fish yields. Parasitic diseases pose a threat to livestock, affecting animal health and productivity, while bacterial diseases can wreak havoc on crops, causing significant economic losses. Additionally, viral diseases can spread rapidly among fish, often with devastating consequences, impacting production.

22

Research-based Shrimp Farming Farm Song Nam

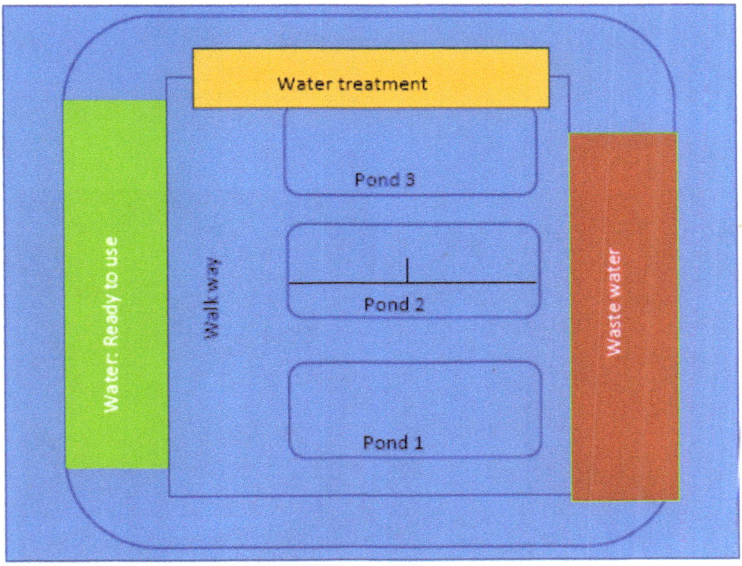

Fig. 59. Diagrammatic presentation of the farm.

Introduction

We, the participants of ITAF 2023 along with Dr. B. R. Chavan and Dr. Prashant Bodake visited Farm Song Nam under the guidance of Asst. Prof. Dr. Naphat Panthukumbol and Dr. Sorawit Powtongsookon 18/10/23 at 01:00 pm. This research station is under NSTDA (National Science and Technology Development Agency). The purpose of the visit was to learn about the latest developments in aquaculture technologies.

IoT-Aqua

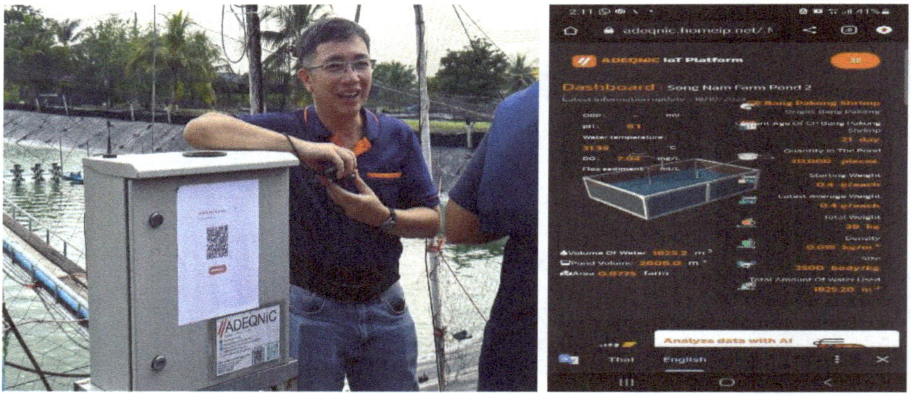

Fig. 60. Dr. Sorawit Powtongsookon explaining about the Aqua-IoT installed at pond 2

Dr. Sorawit explains that this system is able to monitor water quality, floc sedimentation, the current age of shrimp, quantity in the pond, total weight, density, size, total amount of water used, pond volume, and volume of water in the pond at a time. This system is equipped with cloud backup that can store all recorded pieces of information. After monitoring, the available information is automatically uploaded to the cloud and accessible through mobile by simply visiting the website. The key feature of Aqua-IoT is the integration of key data – physical, chemical, and biological water qualities and weather – into one single dashboard that enables users to understand the relationship between the data, analyze the data, and make informed decisions.

Automatic sampler

Fig. 61. The automatic shrimp sampler in a shrimp pond.

This plays a pivotal role in the efficient and accurate monitoring of shrimp populations. This sophisticated tool is designed to collect representative samples of the shrimp population for various purposes, such as assessing growth rates, health, and overall population dynamics. Using a combination of advanced sensors and precision robotics, the sampler can extract shrimp samples at regular intervals, ensuring data accuracy and reducing the need for manual labor. The collected data is instrumental in optimizing feeding strategies, disease management, and overall farm productivity, making the automatic shrimp sampler an indispensable component in modern aquaculture practices. The automatic shrimp sampler in a pond, connected to the cloud and operated through a mobile device from anywhere at any time, represents a remarkable technological advancement in aquaculture.

Costing

Aqua-IoT technology has already been transferred to companies, making the devices commercially available. The system costs around THB 200,000 per pond. Its benefits include cost saving on energy and feed, as well as disease risk reduction. Return on the investment can be expected on the first crop.

Fig. 62. Activities at research based shrimp farm

Under the Leadership

Dr. Sanjay Bhave
Hon. Vice-Chancellor, Dr. B. S. Konkan Krishi Vidyapeeth, Dapoli

Under the Guidance

Dr. Prashant Bodake
Principal Investigator, NAHEP-IDP and
Dean & Director of Instruction, Dr. B. S. Konkan Krishi Vidyapeeth, Dapoli

Co-Principal Investigator, NAHEP-IDP

Dr. Santosh Sawardekar
Dr. V. K. Patil
Dr. Atul Mohad
Dr. R.T. Thokal

Edited and complied by

Dr. B. R. Chavan
Dr. Prashant Bodake
Dr. Manoj Kamble
Dr. Nopadon Pirarat
Dr. Channarong Rodkhum
Dr. A.S. Pawase
Dr. P. E. Shingare
Dr. Seema Medhe
Dr. Aranya Ponpornpisit

Editorial Committee

Amrendra Kumar
Devaranjani VK
Rohit Apte
Gargi Palekar
Viraj Daunde
Swapnil Jadhav
Prachi Dongarkar

Publisher

Dr. Prashant Bodake
Principal Investigator, NAHEP-IDP and
Dean & Director of Instruction, Dr. B. S. Konkan Krishi Vidyapeeth, Dapoli

Dr. Nopadon Pirarat
Professor and Training Organizer,
Chulalongkorn University, Bangkok, Thailand

Financial Support
ICAR-National Agriculture Higher Education Project-Institutional Development Plan (NAHEP-IDP), Indian Council of Agriculture Research, New Delhi

In Collaboration with
Chulalongkorn University, Bangkok, Thailand
Dr. Balasaheb Sawant Konkan Krishi Vidyapeeth, Dapoli, India

Financial Support

ICAR-National Agriculture Higher Education Project-Institutional Development Plan (NAHEP-IDP), Indian Council of Agriculture Research, New Delhi

Brief Description

Book on Water Quality and Fish Health Management will be published during the international conference on International Conference for Implementation of NEP 2020 In Agriculture Universities in Maharashtra Organised by Dr. Balasaheb Sawant Konkan Krishi Vidyapeeth, Dapoli, India. The first section of this book contains hands-on examination of finfish and shellfish Health, includes: Health examination of finfish, necropsy technique and sample collection technique, Microbiology diagnosis, Bacterial isolation and identification, Preparation of subculture plate, Health Examination of Shrimp, Molecular technology. In the second section of this contains water quality of aquaculture pond, it includes, Estimation of pH of aquaculture pond water, Analysis of Ammonia nitrogen, nitrite nitrogen and dissolved oxygen of aquaculture pond water. In the third section of this book contains, small scale aquaculture systems employed in Thailand. This book will be serving as the innovative academic guide line for the students, faculty and farmers to enhance fish production and provide food security and improve their livelihood.

B. R. Chavan

Pre-departure Orientation Program

The pre-departure orientation program for the international training on Aquaculture and Fisheries, organized by NAHEP-IDP, took place at D. B. S. K. K. V. Dapoli. The ceremony was graced by the presence of esteemed dignitaries, including Hon. Vice-Chancellor, Dr. S. G. Bhave, leader, ICAR-NAHEP-IDP, Dr. Balasaheb Sawant Konkan Krishi Vidyapeeth, Dapoli; and Dr. P. S. Bodake, Director of Instruction, and Principal Investigator, ICAR-NAHEP-IDP.

The ceremony commenced with introductory remarks by given by Dr. P. S. Bodake, he informed the significance of the international training and its contribution to the field of Aquaculture and Fisheries. The Participants was welcomed with enthusiasm, acknowledging the academic and practical implications of the upcoming training.

Dr. A. S. Pawase, and Dr. B. R. Chavan, expert in the field of Aquaculture and Fisheries, shared their insights and expectations for the participating students. Their remarks provided valuable context and motivation, emphasizing the importance of this international exposure in shaping the future leaders of the industry.

Following the faculty's remarks, selected students, Amrendra Kumar, Viraj Daunde, Rohit Apte, Devaranjini V. K., Prachi Dongarkar, Swapnil Jadhav and Gargi Palekar had the opportunity to share their thoughts through short speeches. We expressed their excitement, gratitude, and aspirations for the training program. Our perspectives added a personal touch to the event, highlighting the diverse backgrounds and aspirations of the participants.

The highlight of the ceremony was the presidential address delivered by Dr. S. G. Bhave, President NAHEP-IDP; the Vice-Chancellor of Dr. Balasaheb Sawant Konkan Krishi Vidyapeeth, Dapoli. In his speech, he emphasized the strategic importance of international collaborations and the role of NAHEP-IDP in fostering excellence in agricultural education. Dr. Bhave, commended the students for their commitment to advancing their knowledge and skills in Aquaculture and Fisheries, underscoring the global significance of these fields.

Hon. Vice-Chancellor Dr. Bhave has extended his best wishes to the students, encouraging them to make the most of the international training opportunity. His address inspired confidence and instilled a sense of purpose among the participants, fostering a strong sense of pride in being part of the NAHEP-IDP initiative. The ceremony concluded with a vote of thanks delivered by students. The students expressed gratitude to the organizers, faculty, and dignitaries for their support and encouragement.

Fig. 63. Pre-departure Orientation program

International Conference: Recent Advances in Sustainable Tilapia Farming

A One-Day International Conference on the topic 'Recent Advances in Sustainable Tilapia Farming was conducted via Online mode on Zoom. Speakers from various institutes across the globe were called to give lectures on respective topics. Dean and Director of Instructions, DBSKKV, Dapoli Prof. (Dr.) Prashant Bodake had given speech as Chef guest.

The first lecture in the morning session began on the topic 'Global Tilapia Aquaculture by Dr. Partho Pratim Debnath. Followed by this lecture Dr. Gabriel Ataguba has given lecture on the topic 'Recent Advances in the sustainable development of tilapia aquaculture: Production to Consumption. Dr. Sorawit Powtongsook has given lecture on 'Biofloc Technology'. Dr. Balasaheb Chavan presented the topic on 'Cage Farming: Productive and efficient systems. Dr. Naftal Gabriel has presented the topic on the utilization of insects as alternative protein source in Tilapia farming'. We got to know about this new and exiting innovation from this particular lecture. Dr. Ei. Lin Ooi has presented the topic on 'Sustainable aquafeed; plant-based alternatives. Assoc. Prof. Channarong Rodkhum presented the topic on 'non-antibiotic approaches for combat Tilapia bacterial pathogens. Followed by Prof. Kim Thompson presented the topic 'Nano vaccines development', Dr. Teerapong Yata presented the topic 'Herbal nanomedicine', the last lecture of the conference was delivered by Dr. Prapansak Srisapoome on the topic 'Probiotics'.

With this, the One-Day Conference, full of new information and knowledge was concluded with Discussion and closing remarks. All the experts spared their valuable time for doubt clearing which helped to clear concepts and gain extra information related to various topics.

Memorandum of Understanding (MOU) Signing Ceremony

Under the guidance of Hon'ble Vice Chancellor of this University Prof. (Dr.) Sanjay Bhave, A memorandum of understanding (MoU) has been signed between Dr. Balasaheb Sawant Konkan Agricultural University and Chulalongkorn University, Thailand for education and research exchange program both faculty and students.

The MoU signing event organized by Chulalongkorn University, Thailand. Under the guidance of Chulalongkorn University President Prof. (Dr.) Bundhit Eua-arporn, on the said MoU signed Dean and Director of Education Prof. (Dr.) Prashant Bodake and the Dean of the Chulalongkorn university Prof. (Dr.) Sanipa Suradhat has signed on the MoU before the chief guest Ms. **Paulomi Tripathi**, Deputy Chief of Mission and Deputy PR to UNESCAP, Office of the Indian Ambassador, Thailand on dated 20/10/2023.

The MoU signing was done in presence of the officer of Chulalongkorn University. Dr. Nopadon, Dr. Channarong, Dr. Aranya, Dr. Manoj, Dr. Seema and Dr Balasaheb Chavan from Dr. Balasaheb Sawant Kokan Agricultural University and students from College of Fisheries, Ratnagiri: Swapnil Jadhav, Prachi Dongre, Devaranjani K, Gargi Palekar, Rohit Apte, Viraj Daunde and Amarendra Kumar.

The MoU between Chulalongkorn University and Dr. Balasaheb Sawant Konkan Agricultural University, opportunities of higher educations and research will be provided to the graduate and post graduate students of Fisheries/Agriculture and allied disciplines of both the university.

Also, research facilities will be available to the students of both the universities to conduct research. With this MoU, the teachers of both the universities will be able to exchange for teaching and research work. During the MoU program, the Dean of Chulalongkorn University, Prof. (Dr.) Sanipa Suradhat promised to provide higher education and research opportunities to our students in future. Also Dr. Balasaheb Sawant Konkan Agricultural University Dean and Director of Education Prof. (Dr.) Prashant Bodake indicated that higher education opportunities will be available in our universities.

The signing event was smoothly conducted in stipulated time followed by a presentation given by Devaranjini V.K. on overview and remarks on the training conducted for a month and a group photo along with all of we participants. The ceremony concluded with vote of thanks.

FSC
www.fsc.org
MIX
Papier | Fördert
gute Waldnutzung
FSC® C083411

Zeitfracht Medien GmbH
Ferdinand-Jühlke-Straße 7
99095 Erfurt, Deutschland
produktsicherheit@kolibri360.de